T0301848

Clinical Insights for Image-Guided Radiotherapy

This book provides a clinical insight into image-guided radiotherapy (radiation therapy) (IGRT) for prostate cancer. It starts by setting the clinical scene, discussing immobilisation and standard IGRT practice and then considering important developments like IGRT with non-ionising radiation, adaptive radiotherapy, particle therapy, margins, hypofractionation, clinical outcomes, artificial intelligence and training.

Good IGRT requires both technical and clinical focus. So, in complement with our first study guide on IGRT, this book now brings together key, clinical insights into IGRT for prostate cancer patients, with a view to helping the professional learn more about "how-to" undertake IGRT for these patients more accurately, effectively and safely, throughout the whole course of a patient's treatment with radiation.

This clinical insight guide will be of interest to newly qualified radiation therapists, therapeutic radiographers, medical dosimetrists, medical physicists, radiotherapy physicists and clinical oncologists. It will also be of use for trainees and can be used alongside continuing competency and clinical training within clinical departments and radiation therapy centres worldwide.

This is the first in a forthcoming series of clinical insights, each tackling a different treatment area. Further areas in the series will be: the head and neck; thorax; breast; pelvis; and brain.

Key Features:

- Internationally applicable, clinically focused, up-to-date and evidence based
- Accompanied by suitable electronic multimedia resources
- Authored by experts with decades of experience of pioneering electronic portal imaging and IGRT in clinical practice,pedagogic research and substantial experience of teaching/supervising students,trainees and qualified therapists/medical physicists at bachelor, postgraduate and doctoral levels

Mike Kirby and **Kerrie-Anne Calder** are well-respected authors and radiotherapy professionals, who have worked in radiotherapy physics/radiotherapy clinical and academic practice for nearly 35 years and 25 years respectively.

Mike Kirby is a Senior Lecturer in Radiotherapy Physics at the University of Liverpool, UK, and an Honorary Lecturer at the University of Manchester, UK. He holds graduate and postgraduate qualifications in medical physics and has in total over 200 books, papers, oral and poster presentations to his name in the field of radiotherapy. Mike holds professional membership of the Institute of Physics and Engineering in Medicine, the American Association of Physicists in Medicine, the American Society for Radiation Oncology, the European Society for Radiotherapy and Oncology and the British Institute of Radiology and is a Fellow of the Higher Education Academy and the British Institute of Radiology in the UK.

Kerrie-Anne Calder is a Lecturer at the University of Liverpool, UK, where she educates undergraduate and postgraduate students in many aspects of radiotherapy with a special interest and role in imaging training. Kerrie-Anne has graduate and postgraduate qualifications in radiotherapy, education and academic practice, is a member of the Society and College of Radiographers and is a Fellow of the Higher Education Academy in the UK. She was a clinical and professional lead in IGRT (on-treatment verification imaging) within the NHS in the UK for over ten years.

Series in Medical Physics and Biomedical Engineering

Series Editors: Kwan-Hoong Ng, E. Russell Ritenour, and Slavik Tabakov

Recent books in the series:

e-Learning in Medical Physics and Engineering: Building Educational Modules with Moodle
Vassilka Tabakova

Diagnostic Radiology Physics with MATLAB®: A Problem-Solving Approach
Johan Helmenkamp, Robert Bujila, Gavin Poludniowski (Eds)

Auto-Segmentation for Radiation Oncology: State of the Art
Jinzhong Yang, Gregory C. Sharp, Mark Gooding

Clinical Nuclear Medicine Physics with MATLAB: A Problem Solving Approach
Maria Lyra Georgosopoulou (Ed)

Handbook of Nuclear Medicine and Molecular Imaging for Physicists – Three Volume Set
Volume I: Instrumentation and Imaging Procedures
Michael Ljungberg (Ed)

Practical Biomedical Signal Analysis Using MATLAB®
Katarzyna J. Blinowska, Jaroslaw Zygierewicz

Handbook of Nuclear Medicine and Molecular Imaging for Physicists – Three Volume Set
Volume II: Modelling, Dosimetry and Radiation Protection
Michael Ljungberg (Ed)

Handbook of Nuclear Medicine and Molecular Imaging for Physicists – Three Volume Set
Volume III: Radiopharmaceuticals and Clinical Applications
Michael Ljungberg (Ed)

Electrical Impedance Tomography: Methods, History and Applications, Second Edition
David Holder and Andy Adler (Eds)

Introduction to Medical Physics
Cornelius Lewis, Stephen Keevil, Anthony Greener, Slavik Tabakov, Renato Padovani

Calculating X-ray Tube Spectra: Analytical and Monte Carlo Approaches
Gavin Poludniowski, Artur Omar, Pedro Andreo

Problems and Solutions in Medical Physics: Radiotherapy Physics
Kwan-Hoong Ng, Ngie Min Ung, Robin Hill

Biomedical Photonics for Diabetes Research
Andrey Dunaev and Valery Tuchin (Eds)

Spectral Multi-detector Computed Tomography (sMDCT): Data Acquisition, Image Formation, Quality Assessment and Contrast Enhancement
Xiangyang Tang

Monte Carlo in Heavy Charged Particle Therapy: New Challenges in Ion Therapy
Pablo Cirrone and Giada Petringa

Radiotheranostics – A Primer for Medical Physicists I: Physics, Chemistry, Biology and Clinical Applications
Cari Borrás and Michael Stabin

Image-guided Focused Ultrasound Therapy: Physics and Clinical Applications
Feng Wu, Gail ter Haar, and Ian Rivens

Clinical Insights for Image-Guided Radiotherapy: Prostate
Mike Kirby and Kerrie-Anne Calder

For more information about this series, please visit: https://www.routledge.com/Series-in-Medical-Physics-and-Biomedical-Engineering/book-series/CHMEPHBIOENG

Clinical Insights for Image-Guided Radiotherapy

Prostate

Mike Kirby
and
Kerrie-Anne Calder

CRC Press is an imprint of the
Taylor & Francis Group, an **informa** business

Designed cover image: the cover image is taken from a VR 3D rendering of the prostate and surrounding critical structures, obtained from a pelvis CT planning scan, when viewed within the Virtual Environment for Radiotherapy Training (VERT) software (Vertual Ltd., Hull, UK). Image courtesy of Dr James Ward and Prof Andy Beavis, Vertual Ltd.

First edition published 2024
by CRC Press
2385 NW Executive Center Drive, Suite 320, Boca Raton FL 33431

and by CRC Press
4 Park Square, Milton Park, Abingdon, Oxon, OX14 4RN

CRC Press is an imprint of Taylor & Francis Group, LLC

Library of Congress Cataloging-in-Publication Data
Names: Kirby, Mike (Lecturer in radiotherapy physics), author. | Calder, Kerrie-Anne, author.
Title: Clinical insights for image-guidance radiation therapy. Prostate / Mike Kirby and Kerrie-Anne Calder.
Other titles: Prostate | Series in medical physics and biomedical engineering.
Description: First edition. | Boca Raton, FL : CRC Press, 2024. | Series: Series in medical physics and biomedical engineering | Includes bibliographical references and index. | Summary: "This book provides a clinical insight into image-guided radiation therapy (IGRT) on a prostate cancer site. It starts by explaining what is needed for pre-treatment imaging, setting the basis and reference points for what then happens during the treatment phase machine to guide the radiation beams and ensure the treatment is delivered precisely and accurately as planned. Both aspects are needed for solid, continuous professional education in IGRT - and therefore each book covers both and forms an evidence-based 'how-to' guide for IGRT so that it can be undertaken accurately, effectively and safely throughout the whole course of a patient's treatment with radiation. This practical 'how to' guide will be of interest to newly qualified radiation therapists, therapeutic radiographers, medical dosimetrists, medical physicists, radiotherapy physicists and clinical oncologists. They will be of use for trainees and can be used alongside continuing competency and clinical training within real clinical departments and radiation therapy centres worldwide. This is the first in a forthcoming series of guides, each tackling a different treatment area. Further areas in the series will be: Head and Neck; Thorax; Breast; Pelvis; and the Brain"-- Provided by publisher.
Identifiers: LCCN 2023056622 | ISBN 9780367507220 (hardback) | ISBN 9780367497422 (paperback) | ISBN 9781003050988 (ebook)
Subjects: MESH: Prostatic Diseases--radiotherapy | Radiotherapy, Image-Guided--methods | Image Processing, Computer-Assisted
Classification: LCC RC874 | NLM WJ 752 | DDC 616.6/50754--dc23/eng/20240603
LC record available at https://lccn.loc.gov/2023056622

ISBN: 9780367507220 (hbk)
ISBN: 9780367497422 (pbk)
ISBN: 9781003050988 (ebk)

DOI: 10.1201/9781003050988

Typeset in Times
by Deanta Global Publishing Services, Chennai, India

Contents

About the Series

The *Series in Medical Physics and Biomedical Engineering* describes the applications of physical sciences, engineering, and mathematics in medicine and clinical research.

The series seeks (but is not restricted to) publications in the following topics:

- Artificial organs
- Assistive technology
- Bioinformatics
- Bioinstrumentation
- Biomaterials
- Biomechanics
- Biomedical engineering
- Clinical engineering
- Imaging
- Implants
- Medical computing and mathematics
- Medical/surgical devices
- Patient monitoring
- Physiological measurement
- Prosthetics
- Radiation protection, health physics, and dosimetry
- Regulatory issues
- Rehabilitation engineering
- Sports medicine
- Systems physiology
- Telemedicine
- Tissue engineering
- Treatment

The *Series in Medical Physics and Biomedical Engineering* is an international series that meets the need for up-to-date texts in this rapidly developing field. Books in the series range in level from introductory graduate textbooks and practical handbooks to more advanced expositions of current research.

The *Series in Medical Physics and Biomedical Engineering* is the official book series of the International Organization for Medical Physics.

The International Organization for Medical Physics

The International Organization for Medical Physics (IOMP) represents over 18,000 medical physicists worldwide and has a membership of 80 national and 6 regional organizations, together with a number of corporate members. Individual medical physicists of all national member organisations are also automatically members.

The mission of IOMP is to advance medical physics practice worldwide by disseminating scientific and technical information, fostering the educational and professional development of medical physics and promoting the highest quality medical physics services for patients.

A World Congress on Medical Physics and Biomedical Engineering is held every three years in cooperation with International Federation for Medical and Biological Engineering (IFMBE) and International Union for Physics and Engineering Sciences in Medicine (IUPESM). A regionally based international conference, the International Congress of Medical Physics (ICMP) is held between world congresses. IOMP also sponsors international conferences, workshops and courses.

The IOMP has several programmes to assist medical physicists in developing countries. The joint IOMP Library Programme supports 75 active libraries in 43 developing countries, and the Used Equipment Programme coordinates equipment donations. The Travel Assistance Programme provides a limited number of grants to enable physicists to attend the world congresses.

IOMP co-sponsors the *Journal of Applied Clinical Medical Physics*. The IOMP publishes, twice a year, an electronic bulletin, *Medical Physics World*. IOMP also publishes e-Zine, an electronic news letter about six times a year. IOMP has an agreement with Taylor & Francis for the publication of the *Medical Physics and Biomedical Engineering* series of textbooks. IOMP members receive a discount.

IOMP collaborates with international organizations, such as the World Health Organisations (WHO), the International Atomic Energy Agency (IAEA) and other international professional bodies such as the International Radiation Protection Association (IRPA) and the International Commission on Radiological Protection (ICRP), to promote the development of medical physics and the safe use of radiation and medical devices.

Guidance on education, training and professional development of medical physicists is issued by IOMP, which is collaborating with other professional organizations in development of a professional certification system for medical physicists that can be implemented on a global basis.

The IOMP website (www.iomp.org) contains information on all the activities of the IOMP, policy statements 1 and 2 and the 'IOMP: Review and Way Forward' which outlines all the activities of IOMP and plans for the future.

Preface

Image-guided radiotherapy (radiation therapy) (IGRT) has revolutionised external beam radiotherapy over the past few decades, enabling more conformal techniques like intensity modulated radiotherapy (IMRT) and volumetric modulated arc therapy (VMAT) to be delivered with greater accuracy, precision and confidence and paving the way for further advancements through adaptive radiotherapy, dose guidance and particle therapy, to name but a few. For image guidance and on-treatment imaging, the earliest examples were portal films and electronic portal imaging devices to verify geometrically patient set-up. The technology and techniques since then have moved on and continue to develop – all to bring better clinical outcomes for our patients. For prostate cancer (PCa) patients, the main subject matter of this book, modern IGRT has enabled the safe and effective application of hypofractionated and now ultrahypofractionated (UHF) treatment regimes. Online adaptive radiotherapy (ART), as it develops, promises to bring with it great opportunities for safely improving clinical effectiveness still further, for PCa and many other patients.

Our first book, *On-treatment Verification Imaging: A Study Guide for IGRT*, was written as a foundational text, designed to follow the education and training needs of pre-registration therapeutic radiographers who will go on to work in radiotherapy departments around the world. It was also intended to be the ideal study aid for those training to be qualified physicists, clinicians and engineers in radiotherapy. Split into three main sections, it roughly mirrored the educational needs for IGRT for the three years of an undergraduate program of study in radiotherapy. But, as we have found since its publication, there are aspects which are applicable to both graduate and postgraduate programmes in all the major disciplines involved internationally. It serves, as one reviewer kindly put it, "as a template for an IGRT training course; a nice little tool for radiation therapists and junior physicists…with its comprehensive review of a range of topics pertaining to the clinical implementation of IGRT."

As those working with IGRT will appreciate, good IGRT must be clinically focussed and site-specific – and so came the inspiration for this book as a follow-on from the first: to develop a series of clinically focussed guides for IGRT for different cancer sites – in a sense, including many of the aspects which simply could not be fitted into the first book. And so we have this first book of *Clinical Insights for Image-Guided Radiotherapy* – focusing on the prostate.

Radiotherapy for prostate patients is an enormous area of research and development, with a vast evidence base. As such, this book is just as the title suggests – an "insight"; very much borne out of the evidence base, examining many of the different subject areas associated with clinical IGRT for PCa. As such, we have brought together chapters that should set the clinical scene and the need for image guidance on treatment; discuss developments and experiences with patient immobilisation; consider what is now standard IGRT practice for many of our PCa patients as a standard of care; consider IGRT with non-ionising radiation methods; examine the fast-developing and exciting area of change response – adaptive radiotherapy – especially with the latest CT- and MR-guided equipment and methods; consider the role of particle therapy and IGRT experiences therein; consider research and experience of margin reduction and refinement; look at some of the papers associated with the latest developments in hypofractionation and particularly UHF; consider the role of artificial intelligence, as we all must; and finally consider the most important aspect, of continuous training of the cancer professional for modern IGRT.

As far as we are aware, this is the first book of its kind to focus clinically and exclusively on IGRT for an individual cancer site. We hope you find it informative and helpful in what is a tremendously exciting and ever-changing landscape of modern medicine – and that, above all, through your own development in the treatment of cancer, it helps you bring more light into the lives of people who are challenged by it. Thank you for all you do for our patients!

A final word on the terms used in this book. It is written from experience of UK practice, but with a global audience in mind – the principles and points of learning are applicable in many countries that practise radiotherapy (radiation therapy), independent of their development profile. When we use the term "radiographer", we mean "therapeutic radiographer" (UK) and "radiation therapist" (RTT); when we use the term "physicist", we mean "radiotherapy physicist" (or "clinical scientist") (UK) and "medical physicist"; when we use the term "clinician", we mean "trainee clinical oncologist" (UK), "physician" and "radiation oncologist".

If you are completing your training and have just started your first professional roles in radiotherapy (radiation therapy), then this book is written for you!

Acknowledgements

From Mike:

- I'd like to thank many friends, family and colleagues who have supported me throughout this project. The endless conversations, cups of coffee, music, meals and words/gestures of encouragement have meant so much – thank you and praise God for all of you in Liverpool, Chester, Manchester, Preston, Ribchester, Blackburn and many other places! This was not possible without all your love and prayers. This is dedicated to Kerrie-Anne, an amazing work colleague who, in working together for so many years, inspiring and teaching our students…is indeed still doing so! She is the catalyst for our writing projects together and always an amazing person to work with. How she does what she does for her work and all her friends and family, I will never know! And to my family: my brother Anthony and his family, and my dearest parents Lorna and Alf – may they rest in peace. They were, and are, the inspiration for our family lives, and why we do what we do…for the benefit of our patients. Thank you all.

From Kerrie-Anne:

- The biggest thanks go to "Team Calder". Jason, I could not do any of the things I set out to do without you by my side; thanks for all your amazing support while I have been doing this work. My girls, Eleanor and Charlotte, all my work is and will forever be for you two. Charlotte, your smiles and cuddles make the world turn – thank you for the regular supply. Thanks for the support from all friends, family and colleagues; it is fair to say that this book could not have been written without the help from University of Liverpool colleagues. Dr Mike, thank you so much for all the encouragement, the tips, treats and advice. Could not and would not have achieved any of this without you. Next?!
  Finally, this book is dedicated to two amazing people: Eleanor, what a year we have had! Your determination and perseverance through recent struggles have shown me how amazing you are. I love you millions, baby girl, keep being you. Daddo, love and miss you every single day, I still have no doubts x

From us both:

- To Raven Canzeri, Raj Baskara, Sarina Gloster, Prof. Andy Beavis and Prof. Helen McNair at Elekta, Varian Medical Systems, MacroMedics, Vertual Ltd. and the Royal Marsden NHS Foundation Trust, respectively, for supplying many of the excellent images used in this book. To all at Taylor & Francis, especially Danny, for helping, supporting, cajoling and guiding us throughout these months! And especially to our colleagues at the University of Liverpool for helping us with time for writing, huge support and putting up with us in so many different ways….thank you!

Authors: the Revd Canon Dr Mike Kirby and Mrs Kerrie-Anne Calder

Mike and Kerrie-Anne each have over 20 years' experience of developing and practicing the original forms of on-treatment image guidance (electronic portal imaging) in clinical practice, helping to write the current national guidance and leading its implementation in the clinic. They have each been teaching it in clinical practice for over 18 years, and for the past 11 years in the academic setting, using blended teaching and learning methods for allied health professionals.

Mike began work in the UK's National Health Service over 35 years ago, as a Radiotherapy Physicist at the Christie Hospital, Manchester, UK. He then helped set up Rosemere Cancer Centre in Preston, UK as deputy Head of Radiotherapy Physics and Consultant Clinical Scientist. His work moved him back to the Christie as Head of Radiotherapy Physics for the satellite radiotherapy centres and he helped to lead their development in Oldham and Salford as part of the Christie Network. His research has been primarily in electronic portal imaging, developing clinical practice, commissioning and technical implementation of radiotherapy technology (especially electronic portal imaging and radiotherapy networks) and, more recently, teaching and learning for radiotherapy education.

Mike has graduate and postgraduate qualifications in Physics and Medical Physics from the Universities of Durham, Birmingham and Manchester, UK. He has professional membership of the Institute of Physics and Engineering in Medicine (IPEM), the American Association of Physicists in Medicine (AAPM), the American Society for Radiation Oncology (ASTRO), the British Institute of Radiology (BIR), the European Society for Radiotherapy and Oncology (ESTRO); he is a Chartered Scientist (CSci), a Fellow of the Higher Education Academy (FHEA) in the UK and a Fellow of the British Institute of Radiology (FBIR). Mike was an expert lecturer for the IAEA for designing and delivering a week-long Regional Training Course on QA of Record and Verify systems in Algiers (2016). As well as a senior lecturer (Radiotherapy Physics) at the University of Liverpool, he is also an Honorary Lecturer (Faculty of Biology, Medicine and Health) at the University of Manchester and a supervisor for the national HSST Programme for Doctoral level training of clinical scientists within the UK.

Mike is also a priest in the Church of England; having trained and studied at Westcott House and the Universities of Cambridge and Cumbria, he holds graduate and postgraduate degrees in Theology. His ministry has been in the Cathedrals of Blackburn, Chester and Liverpool (Anglican), where he is presently Canon Scientist.

Kerrie-Anne began work in radiotherapy at Clatterbridge Cancer Centre 25 years ago. Following a period of work in Australia, she returned to Clatterbridge. While there, she worked as an imaging specialist radiographer, implementing and overseeing all aspects of the imaging protocol in place. She was also involved in the education of radiographers and students in all aspects of radiotherapy imaging.

Kerrie-Anne currently works at the University of Liverpool, where she educates undergraduate and postgraduate students in many aspects of radiotherapy, with a special interest and role in imaging training.

Kerrie-Anne has graduate qualifications in radiotherapy and oncology from the Universities of Derby and Sheffield Hallam, as well as postgraduate qualifications in teaching and academic practice from the University of Liverpool.

1 Setting the Scene

The aim of this book is to discuss and highlight issues specific to radiotherapy delivered to the prostate.

Safe and effective radiotherapy depends on the accuracy of field placement relative to the target volume. When treating the prostate, consideration must be given to the surrounding healthy tissues and organs. The close proximity of the bladder and bowel to the prostate can lead to these organs being dose-limiting structures. It is vital to ensure that an adequate dose is delivered to the prostate to facilitate a high tumour control probability (TCP) while minimising the normal tissue complication probability (NTCP) or side effects (RCR 2021) (Figure 1.1).

The image presented is of a male pelvis, showing the position of the prostate. Owing to the different procedures that may be involved during gender reassignment, we have not included images related to a female pelvis with a prostate in place. This work, however, does not discriminate between sexes and can be applied to any patient receiving radiotherapy to the prostate.

An added difficulty in ensuring accurate treatment of the prostate is the ability of the gland to move independently of surrounding structures, including the bony anatomy. This means that pelvic bony anatomy cannot be used as a position surrogate for the prostate, limiting the imaging modalities that are suitable for prostate verification. If planar megavoltage (MV) or kilovoltage (kV) imaging is the only available modality, the use of suitable, implanted fiducial markers should be considered. Volumetric imaging is capable of allowing visualisation of the prostate itself and therefore does not require the use of a surrogate such as fiducial markers. There are many advantages and limitations associated with the use of either volumetric imaging or planar imaging with a surrogate. Both modalities will give a snapshot image of the prostate position, generally taken prior to treatment delivery. Standard on-treatment imaging will identify the position of the prostate only before treatment commences. Owing to the likelihood of movement in the pelvis, tracking systems have been developed and used to monitor the prostate position throughout the full treatment fraction delivery; tracking systems include equipment such as implanted transponders and ultrasound. Target tracking can be particularly beneficial during longer treatment delivery times or treatments delivering a high dose per fraction: Stereotactic ablative radiotherapy (SABR) treatment delivery

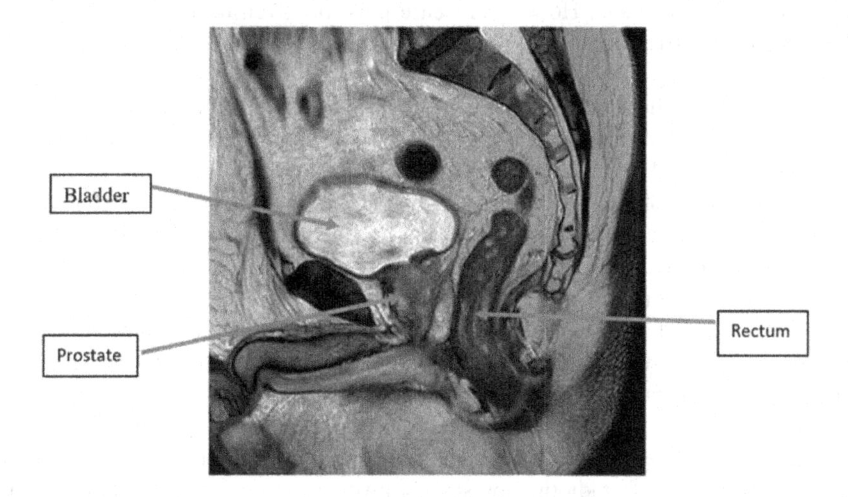

FIGURE 1.1 Sagittal magnetic resonance (MR) pelvis image showing the close proximity of the prostate, rectum and bladder. Image courtesy of Dr Pete Bridge, University of Liverpool.

DOI: 10.1201/9781003050988-1

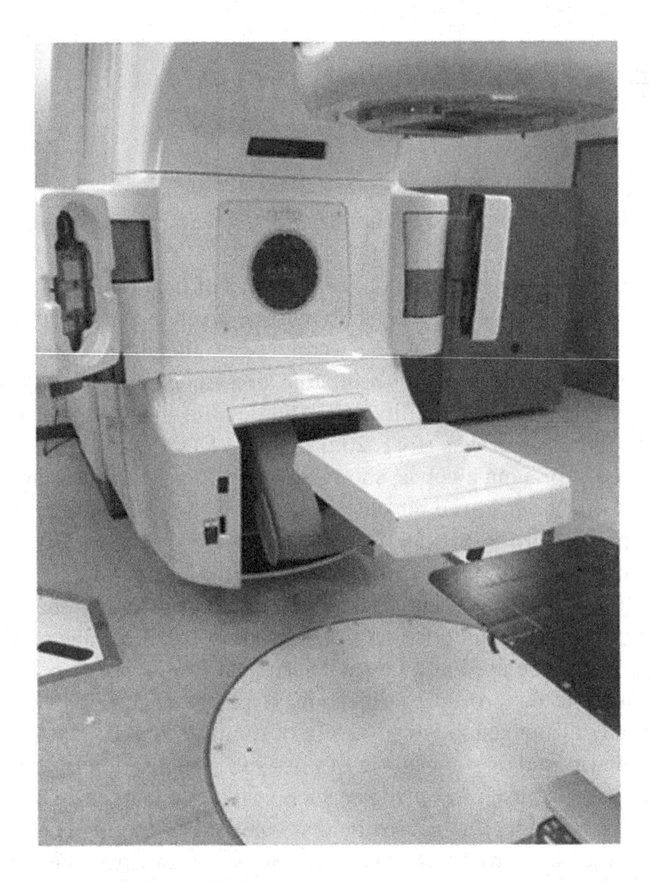

FIGURE 1.2 Standard LINAC (Linear Accelerator) showing MV imaging panel.

involves both an increased time to deliver and a high dose per fraction. Because of the high precision required for SABR treatments owing to the very small margins involved, tracking systems can be required (Figures 1.2–1.5).

If less consideration were given to side effects and the effect of treatment on the surrounding healthy tissue, treatment margins could be increased, and the mobility of the prostate would not pose an issue for target coverage. However, as care must be given to minimise NTCP, treatment margins have been and will continue to be reduced to irradiate as small an amount of healthy tissue as possible.

Radiotherapy treatment has become more conformal to the shape of the target area, with as small a margin applied around the actual tumour or target outline as possible. This is the planning target volume (PTV) that is added to all treatment plans to allow for uncertainties in the plan, including, to some extent, movement of the target volume. This margin is to be large enough to ensure target coverage while also being as small as possible to avoid healthy tissue receiving a high radiation dose. As the rectum and bladder are very close to the prostate, this is a continual challenge in prostate treatment planning (Figure 1.6).

As margins are reduced, the stability of the target area becomes more important. If the target is known to be mobile, as with the prostate, it is important to maintain coverage by increasing margins, monitoring and tracking the target motion or increasing the stability of the target using immobilisation techniques. Some form of immobilisation device is required for all radiotherapy treatments, and prostate treatments require additional immobilisation. External immobilisation devices, commonly used in all radiotherapy setups, involve the use of external equipment placed onto the treatment couch to ensure a reproducible position for patients and to limit patient movement during treatment. However, for patients undergoing prostate therapy, external immobilisation is not

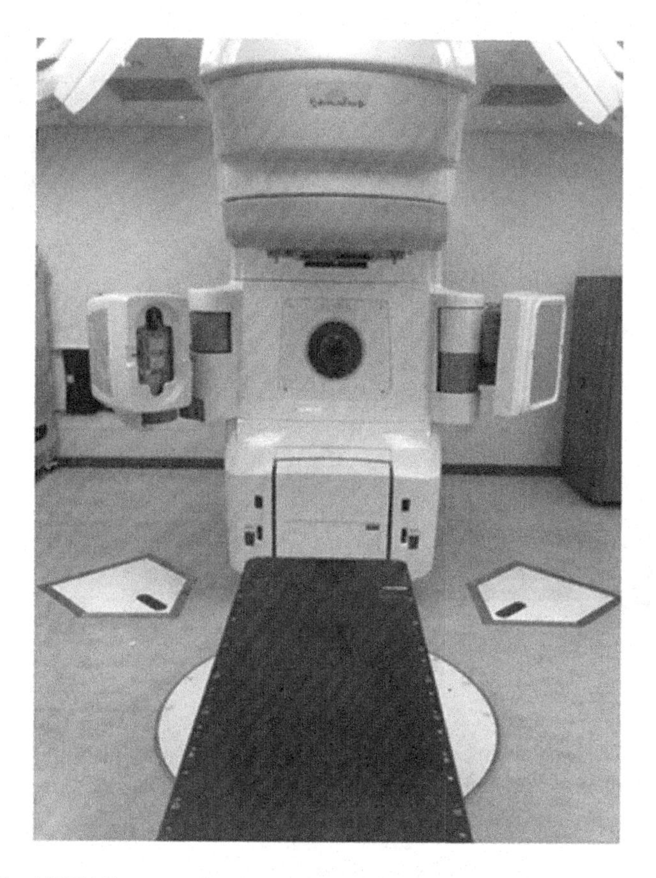

FIGURE 1.3 Standard LINAC gantry showing the kV imaging equipment.

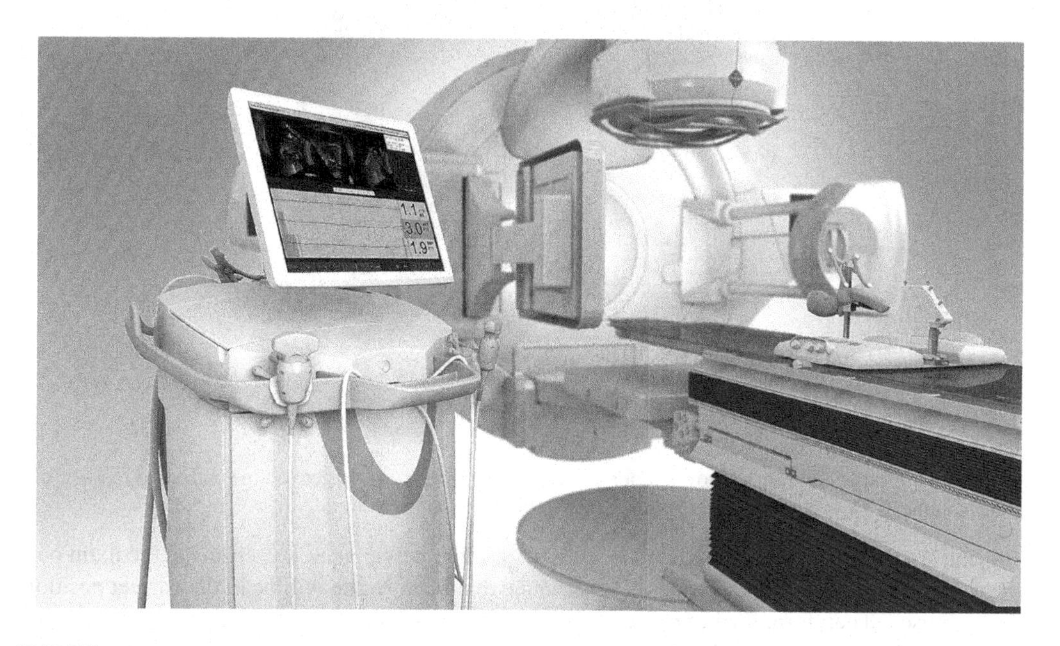

FIGURE 1.4 Elekta Clarity Autoscan ultrasound equipment for identifying and tracking prostate movement. Image courtesy of Elekta.

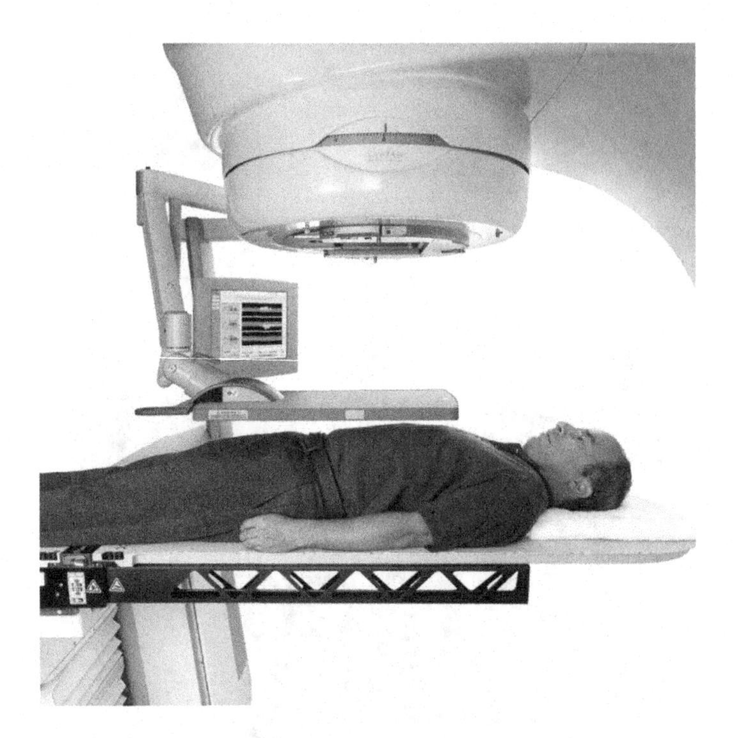

FIGURE 1.5 Varian Calypso tracking system, showing real-time monitoring of implanted transponders in the prostate. Image courtesy of Varian Medical Systems.

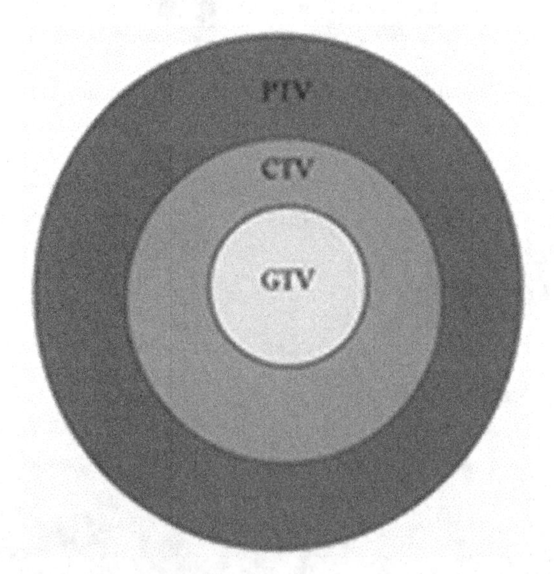

FIGURE 1.6 Simple diagram to show a representation of the three main target volume outlines: GTV – gross tumour volume; CTV – clinical target volume; PTV – planning target volume.

sufficient. As the prostate can move independently of other pelvic anatomy, ensuring the main pelvic bone anatomy is immobilised does not guarantee that the prostate will be in the correct position or immobilised. Therefore, further processes are required, such as a full bladder, rectal balloons, or prostate-rectal spacers (Figure 1.7).

Although efforts are made to limit the movement of the prostate and identify movement through standard online imaging or tracking processes, pelvic anatomy may still differ interfractionally.

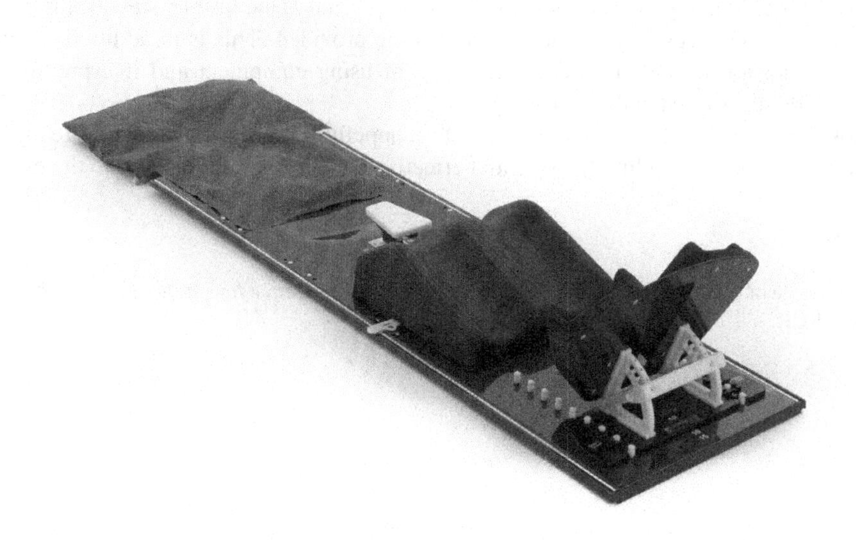

FIGURE 1.7 An example of an external immobilisation device, including alpha cradle, knee support and foot support. Source: © MacroMedics BV.

Adaptive radiotherapy (ART) involves the use of volumetric imaging to assess any anatomical deformity and positional changes prior to each treatment fraction and allows for the plan to be adapted or a replan to take place before treatment. ART can be performed using cone-beam computed tomography (CBCT) or an MR-LINAC. The use of the MR-LINAC will result in no additional imaging dose being delivered to the patient; this has obvious benefits.

Because the aim of maximising dose to the prostate while limiting the dose received by the surrounding organs proves problematic owing to the close proximity of the surrounding organs, the use of proton therapy may be of benefit to prostate patients. As there is no exit dose when using a proton beam, the surrounding organs will receive less or no dose when using proton therapy (Figure 1.8).

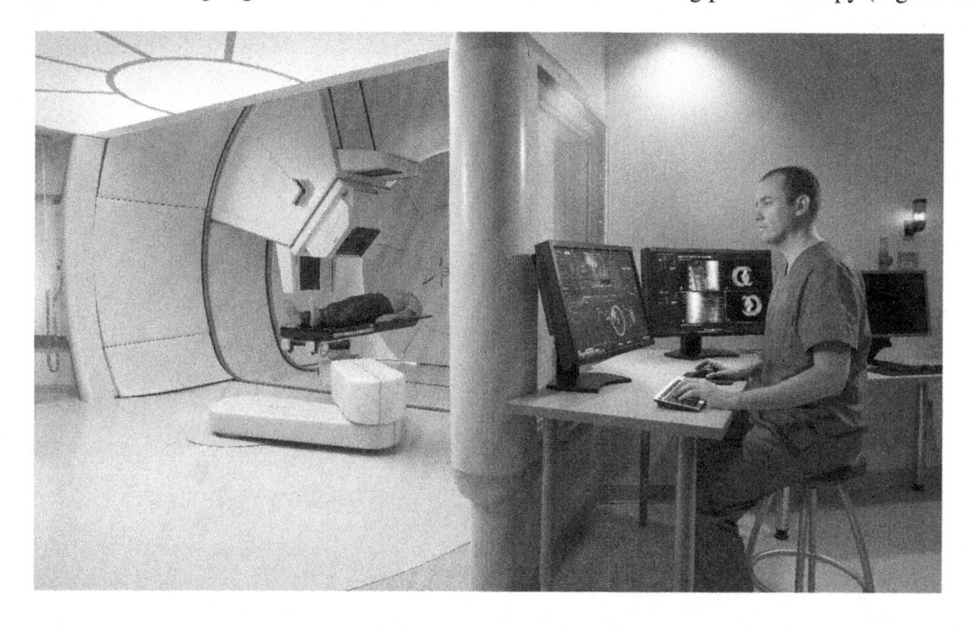

FIGURE 1.8 The Varian ProBeam proton therapy machine. Image courtesy of Varian Medical Systems.

There are many issues to be considered when deciding on the most appropriate method of immobilisation, planning, imaging and treatment of prostate patients. Due to this, staff training around awareness and consequences of these decisions must be provided. This is in addition to the need to provide training for all staff in the practicalities of using equipment and interpreting results obtained from the imaging modalities used.

This book aims to cover the above issues in detail, hopefully leaving the reader more aware and informed of the issues surrounding the safe and effective delivery of radiotherapy to the prostate.

REFERENCE

RCR (Royal College of Radiologists). 2021. ***On target 2: Updated guidance for image-guided radiotherapy***. London: RCR.

2 Prostate Immobilisation

2.1 INTRODUCTION

Radiotherapy to the prostate, as with other structures, has become more conformal as technology has allowed for advances in treatment techniques. The introduction of the multi-leaf collimator (MLC), intensity modulated radiotherapy (IMRT) and more recently volumetric modulated arc therapy (VMAT) and the use of smaller margins has increased the importance and necessity of immobilisation (White et al. 2014). Mobility considerations must not only be given to external anatomy but also to the target itself and the surrounding organs and tissues; prostate position can vary depending on bladder and bowel state (RCR 2021; Gurjar et al. 2020; Jeong et al. 2016).

Due to the internal movement of the prostate gland, independent of the surrounding bony anatomy, both external immobilisation and consideration of internal anatomy mobility should be included in any radiotherapy treatment decisions, specifically with respect to the margins to be used (RCR 2021). External immobilisation will involve the use of devices located on the treatment couch, with the main aim of reducing or preventing gross, conscious movements made by the patient. These devices will also ensure patient position is reproducible between treatment fractions, increasing interfraction reproducibility.

The inclusion of measures to restrict internal motion will aim to stabilise the position of the prostate itself with a view to minimising dosimetric variability throughout the target and surrounding organs during treatment. With regard to minimising the movement of the prostate, the state of the bladder and rectum should be considered (Gurjar et al. 2020; Yahya et al. 2013). The position, size and state of the bladder and bowel can alter the position of the prostate and impact the dose received by the target. There are many ways to achieve internal immobilisation, the more common of which will be discussed further in this chapter. It should be clear that internal immobilisation methods may involve an invasive procedure for the patient and therefore consent and patient compliance should be fully considered before a treatment rationale is finalised. Patient comfort and compliance may be limiting factors in the use of some immobilisation methods discussed (RCR 2021; NRIG 2012).

2.2 EXTERNAL IMMOBILISATION

There are many examples of external immobilisation devices that can be used for pelvis radiotherapy; these include ankle/foot stocks and knee supports, which tend to be produced in a standard form and used for multiple patients. The individualised immobilisation devices include personal vacuum-formed bags and thermoplastic casts. As discussed in the Royal College of Radiologists publication, On-target 2, patient comfort and compliance must be considered when assessing immobilisation devices (RCR 2021).

2.2.1 Foot and Knee Supports

Supports placed under the patient's knees are commonly used to increase patient comfort and compliance with treatment position. These supports can also pay a role in patient immobilisation. The use of a knee support can be used to reduce roll, pitch and yaw in the pelvis and thus aid reproducibility (Kuga et al. 2022; Baumert et al. 2002). The use of a knee support was also found to enable a decrease in margins used and a lowering of the dose received by the bladder and bowel (Kuga et al. 2022). An interesting finding of the study by Kuga et al. was related to patients relaxing as they

DOI: 10.1201/9781003050988-2

FIGURE 2.1 Knee supports, used for comfort or immobilisation, in conjunction with foot supports for pelvis patients. Source © MacroMedics BV.

FIGURE 2.2 Foot supports, used to aid reproducibility of pelvis positioning. Source © MacroMedics BV.

became more used to and aware of the treatment process. This relaxation caused a posterior shift in anatomy in patients treated without knee supports, whereas patients treated in a knee flexion position, due to the support used, did not show this movement, further supporting the use of this immobilisation support device (Kuga et al. 2022; Steenbakkers et al. 2004).

External immobilisation devices are generally used alongside couch top indexing. External immobilisation devices, combined with treatment couch indexing, can be used to improve the accurate positioning of the patient on the treatment couch, while the use of indexing and predetermined treatment couch positions can be used to minimise treatment setup errors, including laterality errors or errors involving incorrect or missed isocentre shifts (Saenz et al. 2018). In addition to this benefit of using couch top indexing, an extra advantage involves ensuring that the immobilisation is routinely located at the correct point for the patient, i.e., the knee supports are located in the same, reproducible place beneath the patient's knees. Incorrect placement of this knee support will alter the angle of the pelvis and may compromise treatment accuracy.

As the images show, without correct positioning of the knee supports, the angle of the patient's pelvis can vary. The use of indexing to locate supports to the treatment couch top, including headrest/pillow, knee supports and ankle/foot stocks, can reduce the potential risk of different pelvic tilts.

The position of permanent marks used for radiotherapy alignment for pelvis patients is generally near to the femoral heads; the skin at this point can be affected by any alteration in joint abduction or adduction. An alteration in the angle of abduction or adduction can be caused by the separation

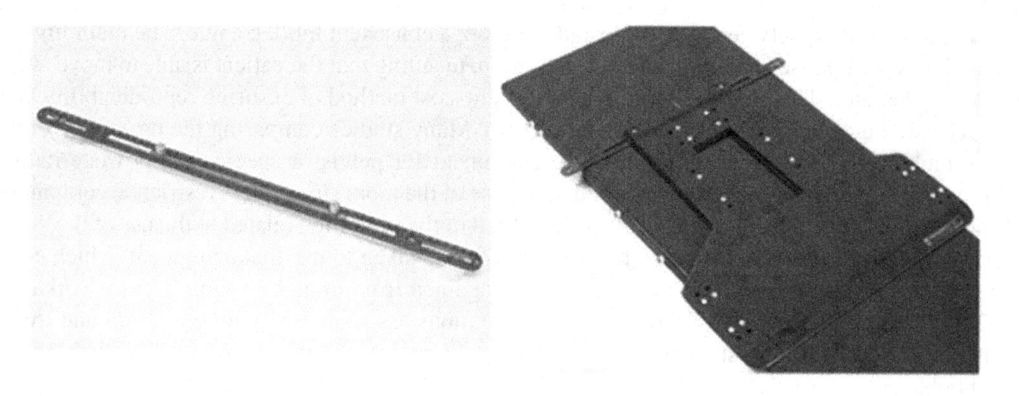

FIGURE 2.3 An indexing locator bar, used to position immobilisation equipment onto the treatment couch in a reproducible position. Source © MacroMedics BV.

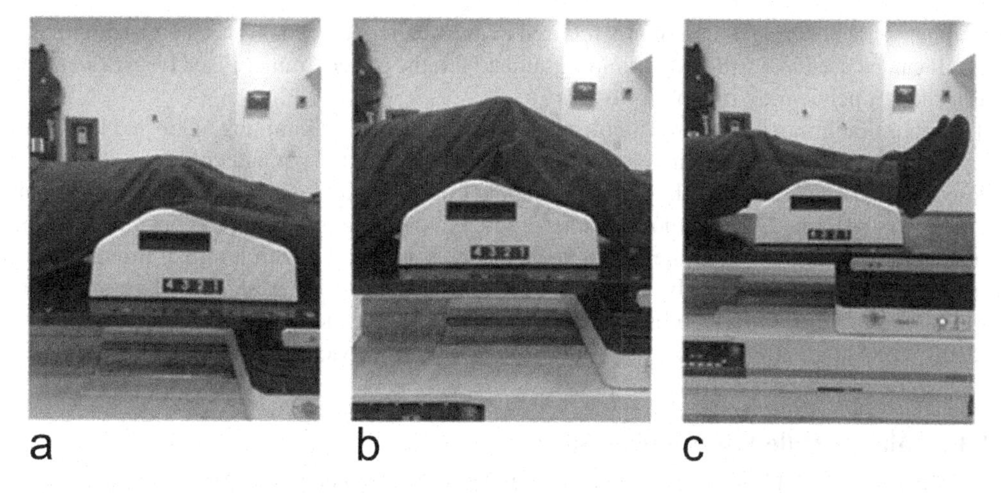

FIGURE 2.4 Images showing the correct position of the knee supports (2.4a), knee supports placed too high (2.4b) or knee supports placed too low (2.4c).

of the patient's legs or ankles on the treatment couch. This issue can be controlled by the use of monitoring or controlling of this separation angle. This can be achieved via a simple measurement of the foot or ankle separation distance while in the treatment position or by the use of immobilisation devices; foot or ankle stocks can be used to keep the patient's legs at a consistent distance apart on the couch. The use of ankle stocks would also ensure that the patient's legs are kept straight and in line with the pelvis and will also provide a level of immobilisation throughout the treatment delivery that the measurement-only technique will not achieve. The use of an ankle measurement is a much simpler, no-cost option and can ensure reproducible leg separation distance. However, it will not immobilise the patient, so that alteration of this distance can occur during set-up or throughout treatment delivery as there is no restriction placed on the patient's movement after set-up. As such, the measurement will require repeated checks and possibly repeat of or alterations to the set-up if it is not correct when checked. This could lead to increased set-up time for each pelvis patient or inaccuracy of leg position during treatment delivery. This measurement technique will also not guarantee that the patient's legs are straight and in line with the patient's pelvis.

2.2.2 Leg Separator

Similar to the ankle or foot stocks, a leg separator can be used. This very simple piece of equipment is of a standard size and is placed between the patient's ankles during set-up. The aim is to ensure a consistent separation between the ankles and therefore a consistent hip/leg angle. The main limitation of leg separator use is the lack of immobilisation, meaning that the patient is able to move away from the separator. The use of this equipment is a low-cost method of ensuring reproducibility and has been well evaluated (Udayashankar et al. 2018). Many studies comparing the use of less rigid leg immobilisation with the more restrictive equipment for pelvis or specifically prostate radiotherapy have reported favourably in relation to the use of the more simple, less restrictive equipment (Udayashankar et al. 2018; Nutting et al. 2000). Much of the limitation related to the use of the more rigid immobilisation is associated with the length of time taken to use that equipment, which does not compensate for the lack of or very slight benefit gained from immobilisation (Udayashankar et al. 2018; Nutting et al. 2000). There are many limitations associated with longer set-up and treatment delivery times for prostate patients, and these will be covered in various sections throughout this book.

2.2.3 Alpha Cradle Immobilisation

The alpha cradle equipment, commonly referred to as a vacuum (vac) bag, is customised immobilisation. It can be used as a full-body immobilisation, with the patient immobilised from head to feet, or as a shorter, pelvis-only immobilisation.

The alpha cradle can be used to combine knee supports while maintaining ankle and leg separation. The manufacture of the equipment and the repositioning of the patient within the alpha cradle is described by Bentel et al. and, although treatment and imaging techniques have evolved since this work was published, the method used for producing the immobilisation has not altered greatly (Bentel et al. 1995). The use of the alpha cradle has been compared with minimal immobilisation devices, such as ankle stocks only (Udayashankar et al. 2018; Nutting et al. 2000; Lee et al. 2014), and also with more rigid immobilisation such as full thermoplastic pelvic casts (Udayashankar et al. 2018; White et al. 2014; Saini et al. 2014). Thermoplastic pelvic casts will be discussed later in this chapter (Subsection 2.2.4).

2.2.3.1 Alpha Cradle Vs Ankle/Foot Stocks

An advantage of the alpha cradle is the use of one piece of equipment, compared with the need for separate knee supports and ankle stocks; these independent pieces of equipment could be positioned incorrectly, limiting the reproducibility of the treatment position. The alpha cradle should offer the ability to immobilise the patient more fully, with the immobilisation involving a greater length of the anatomy; alpha cradles can extend from the patient's head to include the entire length of the

FIGURE 2.5 Material used to produce individualised alpha cradles. Source © MacroMedics BV.

patient, finishing at the feet. They may also be used to include the treatment region only, namely the pelvis in the current study. The aim of the alpha cradle is to ensure that the patient's rotation and overall position achieves a high level of reproducibility and a decrease in movement throughout treatment. Rigid leg immobilisation is believed to aid reproducibility more than pelvis immobilisation only (White et al. 2014). This is suggested to be due to improved patient comfort offered by the custom-formed alpha cradle extending to the patient's feet, and is therefore suggested to be more beneficial than an alpha cradle used for only pelvis immobilisation or the use of less rigid leg immobilisation, such as ankle stocks (White et al. 2014).

When comparing treatment placement accuracy between an alpha cradle and ankle stocks, improvement in the former has been reported in systematic rotational errors (Nutting et al. 2000), although the overall accuracy of treatment position was not improved when using the alpha cradle compared with the less rigid ankle supports. Issues with the method of producing and maintaining the alpha cradle have been reported, resulting in the alpha cradle becoming compromised (Udayashankar et al. 2018). Reproducibility of the treatment position can be reduced when using a vacuum-formed alpha cradle as opposed to a chemically produced one; this is attributed to the loss of vacuum and therefore rigidity of the cradle during prolonged use (Udayashankar et al. 2018). Although loss of vacuum can be seen as an issue when using vacuum-formed immobilisation equipment, Inui et al. acknowledged the issue of "air leakage" in vacuum immobilisation, specifically the latter treatment fractions, but concluded that no effect was seen on the reproducibility of treatment position compared with a non-vacuum-formed alpha cradle (Inui et al. 2018).

The use of an alpha cradle to achieve greater accuracu of treatment placement can be debated and it is not specifically recommended in national guidance (RCR 2021). As improvement has been shown with respect to rotational errors and reproducibility, if rotation is a specific issue for a patient, the use of an alpha cradle could be considered.

Care should be taken when using an alpha cradle for immobilisation as there can be an increase in surface dose received by the patient when treatment fields are planned to pass through the immobilisation device. Jabbari et al. reported an increase in surface dose of up to 30% due to the use of an alpha cradle in the beam path (Jabbari et al. 2018).

2.2.4 Thermoplastic Casts

Thermoplastic casts are commonly used in radiotherapy for effective head and neck or brain immobilisation. They are custom made for each patient and used as a rigid immobilisation of the treatment area. The thermoplastic cast is the only piece of equipment discussed that will restrict external free movement by the patient. Other equipment discussed will limit movement (alpha cradle) or will be used for reproducibility only (leg separator, ankle stocks); these latter immobilisation tools will not restrict a patient's free movement during treatment delivery.

Although the purpose of thermoplastic casts is to restrict and limit movement of the target area, achieving this in the pelvic region is not always seen. There is some disagreement in the literature regarding the use of rigid immobilisation to improve the accuracy of pelvic radiotherapy. Those not reporting an advantage (Udayashankar et al. 2018; Anand et al. 2020) found no benefit of a rigid, thermoplastic pelvic cast over the use of less rigid ankle and foot immobilisation devices, whereas others reported in favour of thermoplastic cast use (Lee et al. 2014; Malone et al. 2000), describing an advantage in reducing mainly pelvic tilt and rotation. The study conducted by Malone et al. reported greater immobilisation when using the thermoplastic equipment compared with alpha cradles and a simple leg cushion; however, it is important to note the different patient positions for this study (Malone et al. 2000). The alpha cradle and the leg cushion were used with patients treated in a supine position whereas the thermoplastic (hipfix) immobilisation was used for patients placed in a prone position. A study by Saini et al. investigated the use of thermoplastic cast arrangements with or without leg support (Saini et al. 2014). Results showed that the use of leg supports improved the accuracy of immobilisation. The leg support was most beneficial when a less immobilising

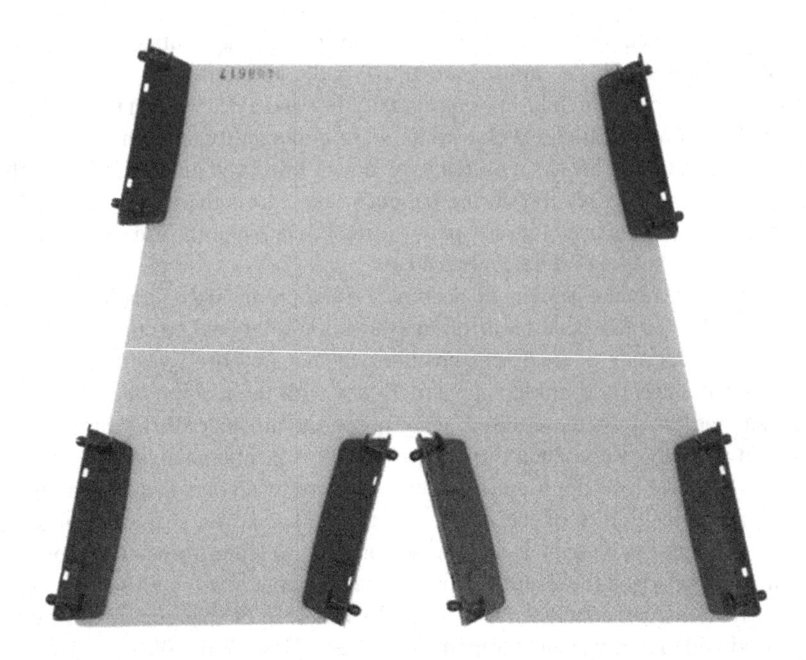

FIGURE 2.6 Thermoplastic material to be formed into a pelvic cast. Source © MacroMedics BV.

cast was used, when the cast was secured using four points and did not cover as much anatomy as the more immobilising cast. This may imply that leg support and separation is the most beneficial aspect of the thermoplastic casts (Saini et al. 2014).

It is important when considering the use of a thermoplastic cast to include a discussion related to the potential increase in the surface dose delivered to the patient. Materials placed onto a patient's surface in the treatment field are known to increase the dose received by the skin (Olch et al. 2014). This increased dose to the skin surface can be minimised by a thinning of the immobilisation material; as the thermoplastic material is stretched around the body, the material thins and has less of an impact on surface dose (Olch et al. 2014). However, as the material is thinned, its ability to be used as a rigid immobilisation device is reduced, as the material loses strength (Hadley et al. 2005). The benefit of a thermoplastic, rigid immobilisation device should be considered alongside the impact of increased surface dose received by the patient. This level of immobilisation may only be beneficial when used alongside leg supports for patients where pelvic rotations are causing concerns for reproducibility and treatment accuracy.

A summary of the main benefits and limitations of each piece of immobilisation device discussed is shown in Table 2.1.

Many of the studies discussed in this section have involved the use of only 2D planar imaging. The limitation of this imaging is the visualisation of only pelvic bony anatomy. We have considered only external immobilisation devices so far. External immobilisation may not have an immobilising effect on the soft tissue within the pelvis; therefore, it is appropriate to include imaging and studies of the bony anatomy only. Although it is important to consider external immobilisation of the pelvis, as has been discussed, the prostate has its own degree of mobility. The prostate is able to move independently of the surrounding bony anatomy and therefore immobilisation of the pelvis as a whole may not result in the immobilisation of the target, i.e., the prostate itself. Internal immobilisation of the prostate must also be considered.

2.3 INTERNAL IMMOBILISATION

As the position of the prostate can be affected by the state of both the bladder and the rectum (Hadley et al. 2005; McNair et al. 2014), ensuring a reproducible state for both of these organs will

TABLE 2.1
A Summary of External Immobilisation Devices

Immobilisation Device	Advantages	Limitations
Ankle separator	Low cost, will not affect patient comfort	Patient not immobilised; no restrictions placed on movement
Ankle/foot stocks	Low cost, reproducible position if used in conjunction with couch indexing, should not cause discomfort	No restrictions placed on movement; leg angle can alter if not indexed to the treatment couch
Leg separator	Low cost, reproducible separation, should not cause discomfort	No restrictions placed on movement, patient's legs could relax away from equipment and increase separation
Vacuum bag	Individualised immobilisation, should improve patient comfort. One piece of equipment minimises incorrect placement of individual pieces	Storage of the vacuum bags can be problematic. loss of immobilisation effect if the vacuum is compromised. Added surface dose must be considered if the treatment beam passes through the vacuum bag
Thermoplastic cast	Reduction of pelvic rotation, rigid immobilisation – patient's movements are restricted, achieving individualised immobilisation	Surface dose is likely to increase as cast covers treatment area; may reduce patient comfort, possible issues if patient has full bladder during treatment

minimise any alteration of the prostate position (RCR 2021). Traditionally during prostate radiotherapy, treatment has been delivered with a full bladder (Tsang et al. 2017), although the need for the bladder to be full for treatment has been debated (Pinkawa et al. 2006). The most important factor is maintaining a consistent bladder volume due to the impact the bladder size has on prostate position (RCR 2021; Gurjar et al. 2020). For treatments involving a full bladder set-up, patients are generally coached in a standard method to ensure a consistent, full bladder state (Nasser et al. 2021). Despite this, there can still be a degree of variability in the bladder size visualised prior to treatment delivery when using a full bladder protocol (Nasser et al. 2021; Byun et al. 2020; Roch et al. 2019). Prostate radiotherapy positioning using an empty bladder protocol has been shown to be non-inferior dosimetrically, compared with using a full bladder protocol (Tsang et al. 2017; Byun et al. 2020), although some concern regarding bladder dose has been discussed (Chetiyawardana et al. 2020). The use of an empty bladder for prostate treatment could be considered a more consistent and stable set-up due to patient comfort and compliance (Tsang et al. 2017), or difficulties in achieving and maintaining a full bladder due to treatment toxicities (Byun et al. 2020).

In a similar way to the bladder, the size and shape of the rectum can affect prostate position, so measures should be included to achieve rectum size consistency (Roch et al. 2019); the standard guidance is to empty the rectum prior to prostate radiotherapy treatment (RCR 2021), although the methods used to achieve this can be varied (RCR 2021; Yahya et al. 2013; McNair et al. 2014). An empty bowel is recommended as it is more consistent in terms of bowel size during treatment; the rectum is known to decrease in size throughout treatment due to irritation caused by acute radiation side effects (RCR 2021). Comparisons of methods for achieving an empty bowel have been researched; comparisons have been made between dietary advice given to the patient to be followed throughout their time under treatment, encouraging the use of high-fibre foods, or the use of laxatives and micro-enemas immediately before treatment delivery (Yahya et al. 2013; Roch et al. 2019). Dietary advice to maintain consistent bowel habits and rectum size for treatment may be seen as non-invasive and potentially the easiest bowel preparation technique to use; there may be issues seen with compliance as treating staff will have no way to monitor or ensure that advice is followed. The use of daily micro-enemas can give a more consistent bowel state as the use of the enema should ensure an empty rectum and patients can be instructed on the timing of use. Patient understanding and compliance are needed with this more invasive procedure and there will remain an element of treating staff being unaware if the procedure has been followed and completed successfully prior to treatment and imaging verification.

2.3.1 Rectal Balloons

For prostate motion management consistency of the bowel size and position is required (RCR 2021), as it is with the bladder; however, the preferred state of the bladder and bowel for treatment can vary and either full or empty states can be used for treatment. Although a full or empty bladder state can be controlled to some degree by the patient, a full and consistent bowel or rectum cannot be managed as easily. Prostate radiotherapy delivered using a full rectum is most commonly achieved with the use of a rectal balloon. For this procedure, a balloon rectal tube is inserted into the patient's rectum daily, prior to treatment and planning scans. The rectal balloon is then inflated with air to a specific size. The balloon will fill the rectum, moving the prostate and anterior rectal wall anteriorly and towards the symphysis pubis. The posterior rectal wall's position will not be changed. In this way, movement of the prostate is restricted (Wachter et al. 2002). This method gives a reproducible rectal state for radiotherapy treatment and will achieve a degree of immobilisation of the prostate (Cho et al. 2009; Smeenk et al. 2010, Smeenk et al. 2012). Another advantage of the use of rectal balloons is the reduction in the dose received by the rectum (Cho et al. 2009; Smeenk et al. 2012; Both et al. 2012). Despite the advantages in the use of rectal balloons, it is an invasive procedure that requires patient consent and compliance. As with any invasive procedure, a certain level of patient discomfort is to be expected, although discomfort when using rectal balloons has not been reported as a limiting factor for use, with initial discomfort resolving quickly (Smeenk et al. 2010).

Further limitation to the use of rectal balloons is the reproducibility achieved when inserting and during inflation (Cho et al. 2009; Smeenk et al. 2010). This issue was discussed by Cho et al., who developed a modified rectal balloon to aid consistent insertion and inflation (Cho et al. 2009). The modified insertion tube included a scale to monitor depth of insertion and a balloon specifically designed to inflate symmetrically, reducing the main concern of reproducibility when using balloons.

As with any material placed in the treatment beam path, the effect on the dosimetry of using rectal balloons should be considered. Srivastava et al. researched the dosimetric effect of the rectal balloon, comparing different balloon-filling materials (Srivastava et al. 2013). While acknowledging the immobilisation effects of the rectal balloon on the prostate and the reduction in rectal dose achieved, there was concern discussed regarding dose non-homogeneity when using air-filled balloons. This discrepancy was reduced when using water or contrast-filled balloons (Srivastava et al. 2013); if using rectal balloons, thought should therefore be given to the filling agent used.

2.3.2 Rectal-prostate Spacers

The location of the rectum, in close proximity to the prostate, can lead to the rectum being a dose-limiting organ (Serrano et al. 2017; Pinkawa et al. 2011). The dose received by the rectum can lead to long-term side effects from radiotherapy and must be kept as low as possible; the area of the rectum receiving a radiation dose must also be limited (Schorghofer et al. 2019; Padmanabhan et al. 2017). A decrease in the dose received by the rectum can be achieved by adequate immobilisation of the prostate, allowing for accurate dose delivery, or by separating the prostate from the rectum. The increase in distance between the rectum and the prostate can be achieved by inserting a spacer between the two organs. The spacer placed between the prostate and rectum can involve the use of a variety of materials, the most common of which is the use of a fluid-filled balloon or the injection of a poly(ethylene glycol) gel (Schorghofer et al. 2019; Wolf et al. 2015). The use of a rectal spacer has been shown to decrease the dose received by the rectum (Pinkawa et al. 2011; Armstrong et al.

2021) and therefore decrease side effects and complications associated with prostate radiotherapy (Padmanabhan et al. 2017; Mok et al. 2014). Spacer use involves an invasive procedure for the patient, and patient consent is required. The use of a spacer involves a surgical procedure; the spacer is inserted prior to radiotherapy planning and remains *in situ* for the duration of the treatment. This should limit the inconsistencies of insertion procedure seen with rectal balloon use (Wolf et al. 2015). Although Wolf et al. discussed the consistency of the gel spacers, there is some concern regarding the spacers maintaining their original dimensions throughout the full treatment course (Wolf et al. 2015). The material used for the spacer will be absorbed into the body after approximately three months, so there is no requirement for the spacer to be surgically removed following completion of radiotherapy; patients will undergo one procedure to insert the spacer rather than daily insertion of a rectal balloon (Padmanabhan et al. 2017). Spacers are generally well tolerated by patients and have not resulted in any patient discomfort, other than that expected following any insertion procedure (Padmanabhan et al. 2017). Schorghofer et al. discussed toxicities related specifically to the use of a spacer; although the study concluded a low overall complication rate, an increase was seen in the risk of possible necrosis or rectal perforations (Schorghofer et al. 2019). This risk should be considered before use. Schorghofer et al. recommended the use of spacers to be limited to treatments involving escalated doses of 80 Gy or greater.

Immobilisation is an important aspect of accurate radiotherapy treatment delivery; accuracy can only be confirmed with the use of adequate on-treatment imaging verification. It is important to visualise the target or a suitable surrogate. External immobilisation used for pelvic treatments is aimed at reducing gross patient movements and ensuring reproducibility of the patient's position on the treatment couch; for this reason, planar imaging may be sufficient. However, as the prostate is able to move independently of the pelvic bones and surrounding anatomy, internal immobilisation is recommended. It is important to ensure that the target position is confirmed and any internal immobilisation is effective; for this reason, imaging of the target using volumetric imaging or the use of a fiducial as a surrogate for the prostate is required to confirm the position of the target prior to treatment delivery. Imaging options will be discussed in the next chapter.

REFERENCES

Anand, M., Parikh, A., Shah, S. P. 2020, April–June. Comparison of thermoplastic masks and knee wedge as immobilisation devices for image-guided pelvic radiation therapy using cone beam computed tomography. *Indian Journal of Cancer.* 57(2): 182–186.

Armstrong, N., Bahl, A., Pinkawa, M., et al. 2021. SpaceOAR hydrogel spacer for reducing radiation toxicity during radiotherapy for prostate cancer. A systemic review. *Urology.* 156: e74–e85

Baumert, B. G., Zagralioglu, O., Davies, B., et al. 2002. The use of a leg holder immobilisation device in 3D-conformal radiation therapy of prostate cancer. *Radiotherapy and Oncology.* 65: 47–52.

Bentel, G. C., Marks, L. B., Sherouse, G. W., et al. 1995. The effectiveness of immobilisation during prostate irradiation. *International Journal of Radiation Oncology, Biology and Physics.* 31(1): 143–148.

Both, S., Deville, C., Bui, V., et al. 2012. Emerging evidence for the role of an endorectal balloon in prostate radiation therapy. *Translational Cancer Research.* 1(3): 227–235.

Byun D. J., Gorovets, D. J., Jacobs, L. M., et al. 2020. Strict bladder filling and rectal emptying during prostate SBRT@ Does it make a dosimetric or clinical difference? *Radiation Oncology.* 15: 239.

Chetiyawardana, G., Hoskin, P. J., Tsang, Y. M. 2020. The implementation of an empty bladder filling protocol for localised prostate volumetric modulated arctherapy (VMAT): Early results of a single institution service evaluation. *British Journal of Radiology.* 93: 20200548.

Cho, J. H., Lee, C-G., Kang, D. R., et al. 2009. Positional reproducibility and effects of a rectal balloon in prostate cancer radiotherapy. *Journal of Korean Medical Science.* 24: 894–903.

Gurjar, O. P., Arya, R., Goyal, H. 2020. A study on prostate movement and dosimetric variation because of bladder and rectum volumes changes during the course of image-guided radiotherapy in prostate cancer. *Prostate International.* 8: 91–97.

Hadley, S. W., Kelly, R., Lam, K. 2005. Effects of immobilisation mask material on surface dose. *Journal of Applied Clinical Medical Physics.* 6(1, Winter): 1–7.

Inui, S., Ohira, S., Isono, M., et al. 2018, October–December. Comparison of interfractional setup reproducibility between two types of patient immobilisation devices in image-guided radiation therapy for prostate cancer. *Journal of Medical Physics*. 43(4): 230–235.

Jabbari, K., Almasi, T., Rostampour, N., et al. 2018, October–December. Evaluating the effect of the vacuum bag on the dose distribution in radiation therapy. *Journal of Cancer Research and Therapeutics*. 14(6).

Jeong, S., Lee, J. H., Chung, M. J., et al. 2016, January. Analysis of geometric shifts and proper set up-margin in prostate cancer patients treated with pelvic intensity-modulated radiotherapy using endorectal ballooning and daily enema for prostate immobilisation. *Medicine*. 95(2): 1245–1250.

Kuga, N., Shirieda, K., Sato, Y., et al. 2022, January–March. Original knee fixation device as a useful fixation method during prostate intensity-modulated radiation therapy. *Journal of Medical Physics*. 47(1): 1–6.

Lee, J. A., Kim, C. Y., Park, Y. J., et al. 2014. Interfractional variability in intensity-modulated radiotherapy of prostate cancer with or without thermoplastic pelvis immobilisation. *Strahlentherapie and Onkologie*. 190: 94–99.

Malone, S., Szanto, J., Perry, G., et al. 2000. A prospective comparison of three systems of patient immobilisation for prostate radiotherapy. *International Journal of Radiation Oncology, Biology and Physics*. 48(3): 657–665.

McNair, H. A., Wedlake, L., Lips, I. M., et al. 2014. A systematic review: Effectiveness of rectal emptying preparation in prostate cancer patients. *Practical Radiation Oncology*. 4: 437–447.

Mok, G., Benz, E., Vallee, J-P., et al. 2014. Optimisation of radiation therapy techniques for prostate cancer with prostate-rectum spacers: A systemic review. *International Journal of Radiation Oncology, Biology and Physics*. 90(2): 278–288.

Nasser, N. J., Fenig, E., Klein, J., Agbarya, A. 2021. Maintaining consistent bladder filling during external beam radiotherapy for prostate cancer. *Technical Innovations and Patient Support In Radiation Oncology*. 17: 1–4.

National Radiotherapy Implementation Group (NRIG). 2012. *Image guided radiotherapy- Guidance for implementation and use*. London: National Cancer Action Team.

Nutting, C. M., Khoo, V. S., Walker, V., et al. 2000. A randomised study of the use of a customised immobilisation system in the treatment of prostate cancer with conformal radiotherapy. *Radiotherapy and Oncology*. 54: 1–9.

Olch, A. J., Lee, G., Heng, L., et al. 2014, June. Dosimetric effects caused by couch tops and immobilisation devices: Report of AAPM task group 176. *Medical Physics*. 41(6): 061501-1–061501-30.

Padmanabhan, R., Pinkawa, M., Song, D. Y. 2017. Hydrogel spacers in prostate radiotherapy: A promising approach to decrease rectal toxicity. *Future Oncology*. 13(29): 2697–2708.

Pinkawa, M., Asadpour, B., Gagel, B., et al. 2006. Prostate position variability and dose-volume histograms in radiotherapy for prostate cancer with full and empty bladder. *International Journal of Radiation Oncology, Biology and Physics*. 64(3): 856–861.

Pinkawa, M., Corral, N. E., Caffaro, M., et al. 2011. Application of a spacer gel to optimise three-dimensional conformal and intensity modulated radiotherapy for prostate cancer. *Radiothrapy and Oncology*. 100: 436–441.

RCR (Royal College of Radiologists). 2021. *On target 2: Updated guidance for image-guided radiotherapy*. London: RCR.

Roch, M., Zapatero, A., Castro, P., et al. 2019. Impact of rectum and bladder anatomy in intrafractional prostate motion during hypofractionated radiation therapy. *Clinical and Translational Oncology*. 21: 607–614.

Saenz, D. L., Astorga, N. R., Kirby, N., et al. 2018. A method to predict patient-specific table coordinates for quality assurance in external beam radiation therapy. *Journal of Applied Clinical Medical Physics*. 19(5): 625–631.

Saini, G., Aggarwal, A., Jafri, S. A., et al. 2014, October–December. A comparison between four immobilisation systems for pelvic radiation therapy using CBCT and paired kilovoltage portals based image-guided radiotherapy. *Journal of Cancer Research and Therapeutics*. 10(4): 932–936.

Schorghofer, A., Drerup, M., Kunit, T., et al. 2019. Rectum- spacer related acute toxicity- endoscopy results of 403 prostate cancer patients after implantation of gel or balloon spacers. *Radiation Oncology*. 14: 47.

Serrano, N. A., Kalman, N. S., Anscher, M. S. 2017. Reducing rectal injury in men receiving prostate cancer radiation therapy: Current perspectives. *Cancer Management And Research*. 9: 339–350.

Smeenk, R. J., Louwe, R. J. W., Langen, K. M., et al. 2012. An endorectal balloon reduces intrafraction prostate motion during radiotherapy. *International Journal of Radiation Oncology, Biology and Physics*. 83(2): 661–669.

Smeenk, R. J., The, B. S., Butler, B., et al. 2010. Is there a role for endorectal balloons in prostate radiotherapy? A systematic review. *Radiotherapy And Oncology*. 95: 277–282.

Srivastava, S. P., Das, I. J., Kumar, A., et al. 2013. Impact of rectal balloon-filling materials on the dosimetry of prostate and organs at risk in photon beam therapy. *Journal of Applied Clinical Medical Physics.* 14(1): 81–91.

Steenbakkers, R. J. H. M., Duppen, J. C., Betgen, A., et al. 2004. Impact of knee support and shape of tabletop on rectum and prostate position. *International Journal of Radiation Oncology, Biology and Physics.* 60(5): 1364–1372.

Tsang Y. M., Hoskin, P. 2017. The impact of bladder preparation protocols on post treatment toxicity in radiotherapy for localised prostate cancer patients. *Technical Innovations and Support in Radiation Oncology.* 3–4: 37–40.

Udayashankar, A. H., Noorjahan, S., Srikantia, N., et al. 2018. Immobilisation versus no immobilisation for pelvic external beam radiotherapy. *Reports of Practical Oncology and Radiotherapy.* 23: 233–241.

Wachter, S., Gerstner, N., Dorner, D., et al. 2002. The influence of a rectal balloon tube as internal immobilisation device on variations of volumes and dose-volume histograms during treatment course of conformal radiotherapy for prostate cancer. *International Journal of Radiation Oncology, Biology and Physics.* 52(1): 91–100.

White, P., Yee, C. K., Shan, L. C., et al. 2014. A comparison of two systems of patient immobilisation for prostate radiotherapy. *Radiation Oncology.* 9:29.

Wolf, F., Gaisberger, C., Ziegler, I., et al. 2015. Comparison of two different rectal spacers in prostate cancer external beam radiotherapy in terms of rectal sparing and volume consistency. *Radiotherapy and Oncology.* 116: 221–225.

Yahya, S., Zarkar, A., Southgate, E., et al. 2013. Which bowel preparation is best? Comparison of a high-fibre diet leaflet, daily microenema and no preparation in prostate cancer patients treated with radical radiotherapy to assess the effect on planned target volume shifts due to rectal distension. *British Institute of Radiology.* 86: 20130457.

3 Prostate Standard Imaging

3.1 INTRODUCTION

Accepted guidance for prostate radiotherapy verification imaging is the use of daily image guidance (image-guided radiotherapy, IGRT) (RCR 2021). It is important to ensure visibility of the prostate's position during treatment due to the fact that it is able to move independently of the surrounding anatomy as well as the use of more conformal treatment techniques leading to the need for greater positional precision (Moseley et al. 2007; O'Neil et al. 2016). The use of planar imaging and a bone match for prostate verification has been shown to be ineffective due to prostate mobility (Crook et al. 1995). Considerations for prostate imaging verification include the use of fiducial markers and planar imaging or, alternatively, volumetric imaging and soft tissue visualisation when using cone-beam computed tomography (CBCT). There has been much research published comparing the use of the two methodologies (Moseley et al. 2007; Das et al. 2014; Deegan et al. 2015). As well as discussions relating directly to using a fiducial marker match compared with a soft tissue match, secondary issues include interoperator variability, concomitant dose delivered and the speed and timing associated with the different imaging modalities. The decision as to which imaging method to use will be influenced by, and will, in turn, influence other aspects of treatment and planning for these patients, such as the immobilisation method used (internal or external) and the margins used at planning; and will be discussed fully in Chapter 7.

3.2 FIDUCIAL MARKERS

Fiducial markers are used as surrogates to visualise or track the location of the treatment target. They are placed inside or adjacent to the target organ or tissue (Chan et al. 2015). It is important that the material used for the fiducial marker is visible using the available imaging equipment and does not produce artefacts that could distort the image and affect its use. For these reasons, gold seeds are used as fiducial markers in prostate radiotherapy (O'Neil et al. 2016; Chan et al. 2015; Das et al. 2014). The gold seeds are inserted into the prostate using transrectal ultrasound to visualise and aid insertion (Ghaffari et al. 2019; Schiffner et al. 2007).

Three seeds are inserted into the prostate before treatment planning takes place. There has been concern reported regarding fiducial marker migration following insertion, questioning the reliability of the surrogates (Ariyaratne et al. 2016; O'Neil et al. 2016). However, there have been studies reporting such migration to have no significant effect on the position of the fiducial markers and that these markers are an accurate method of locating the prostate (Schiffner et al. 2007; Kupelian et al. 2005). Migration should be considered with image registration, using the centre mass of the implanted markers rather than the position of each individual gold seed (Kupelian et al. 2005; Delouya et al. 2010). A suggested method to reduce the impact of migration is the timing of insertion before planning scans are completed (O'Neil et al. 2016; Delouya et al. 2010). Delouya et al. suggested leaving an adequate time (three days) to allow for any post-insertion swelling or oedema to resolve before a planning scan is taken; fiducial insertion and planning scan on the same day is not recommended (Delouya et al. 2010). Changes in position of the fiducial markers can be seen throughout the duration of radiotherapy and is considered to be due more to a change in prostate volume than to fiducial marker migration (O'Neil et al. 2016; Kupelian et al. 2005). Prostate volume may change throughout treatment due to oedema caused either by the insertion of the fiducial markers or the radiotherapy reactions, with either leading to an initial increase in volume (Van der Heide et al. 2007, King et al. 2011). The measured prostate volume, following the initial increase, can be seen to decrease as treatment continues, due to the resolution of the initial swelling (Van der Heide

DOI: 10.1201/9781003050988-3

et al. 2007) or to hormone therapy causing a decrease in prostate volume (McNair et al. 2008). A limitation of the use of fiducial markers is the inability to observe and measure the prostate volume changes throughout treatment as only the fiducial markers and their position are visible on planar imaging (Delouya et al. 2010; Eren et al. 2020). Volumetric imaging with the markers in place would be needed to assess the position of the fiducials and measure the prostate volume.

One of the main causes of fiducial marker movement is the effect of the surrounding tissues on the prostate position or shape (O'Neil et al. 2016; Van der Heide et al. 2007). Bladder and bowel state can, as previously discussed in Chapter 2, cause an alteration in the position of the prostate gland (Munoz et al. 2012; Balter et al. 1995). The study by Balter et al., although now quite dated, shows that the issue of prostate movement and the effect of bladder and bowel state on this movement have been identified and studied for a number of years. It is only as imaging technology and immobilisation equipment have developed that we are now able to account for and minimise the impact of this movement. It is important when using fiducial markers to consider the difference between prostate movement due to surrounding tissues and organs and the potential for marker migration. Marker migration can be assessed by measuring the distance between the markers taken on daily images (Delouya et al. 2010; McNair et al. 2008). Marker migration studies have shown that fiducial migration is minimal and does not limit the use of fiducials as a surrogate for prostate position (Kupelian et al. 2005; Van der Heide et al. 2007).

Insertion of fiducial markers requires a surgical procedure, generally under only a local anaesthetic and needing full patient consent and compliance. Insertion is generally well tolerated (Brown et al. 2020; McNair et al. 2008), with few patients reporting complications. The possible issues that may arise from insertion include infection of the insertion site, pain and prostate deformation (Van der Weilen et al. 2008, Loh et al. 2015; Deegan et al. 2015). Insertion can be achieved via two routes, namely transrectal or transperineal. Both routes are well tolerated but, due to the risk of infection following a surgical procedure, prophylactic antibiotics are given, with an increased dose or change of medication should the patient develop an infection (Loh et al. 2015). Investigations into infection rates following fiducial insertion have been reported as low but require monitoring (Loh et al. 2015), and do not limit the use of prostate fiducial markers (Ghaffari et al. 2019).

As discussed in the immobilisation section of Chapter 2, it is important to have internal immobilisation of the prostate as well as external immobilisation of the patient. Internal immobilisation involves a method of maintaining the bladder and bowel state. Some methods are more invasive than others: rectal balloons, for example, are more invasive for the patient than treatment with a self-regulated empty bowel. Many internal immobilisation methods will cause a level of discomfort for the patient, such as rectal balloons and a full bladder for example, so it is important to consider the time taken for on-treatment imaging to be completed, as this will add to the patient's overall time in the treatment position (Steiner et al. 2013). Patients who are uncomfortable in the treatment position are more likely to move during treatment delivery. If the patient does move out of the treatment position following verification images, they will require repositioning, and any images taken will need to be repeated to ensure the accuracy of the new set-up position. This will increase the dose given to the patient and increase the patient's in-room time. Therefore, it is essential that the time taken for imaging is considered (Webster et al. 2020; Ghaffari et al. 2019). The time taken includes both the time to acquire the image and the time to analyse and make decisions based on that image. Fiducial marker use is generally associated with planar, orthogonal images. In addition to delivering a lower radiation dose to the patient compared with volumetric imaging, this imaging modality takes less time to acquire an image than does a cone beam CT image since the LINAC (Linear Accelerator) is not required to complete a full arc to produce the image. Image registration or matching for fiducial markers is a simple task as the only focus of the registration is the position of the gold seed fiducials. No consideration of soft tissue or bony anatomy is included in the match as they are generally not visible, either due to the modality used (i.e., planar images) or the image field size not including bone anatomy.

A typical image pair using fiducials for registration is shown in Figure 3.1.

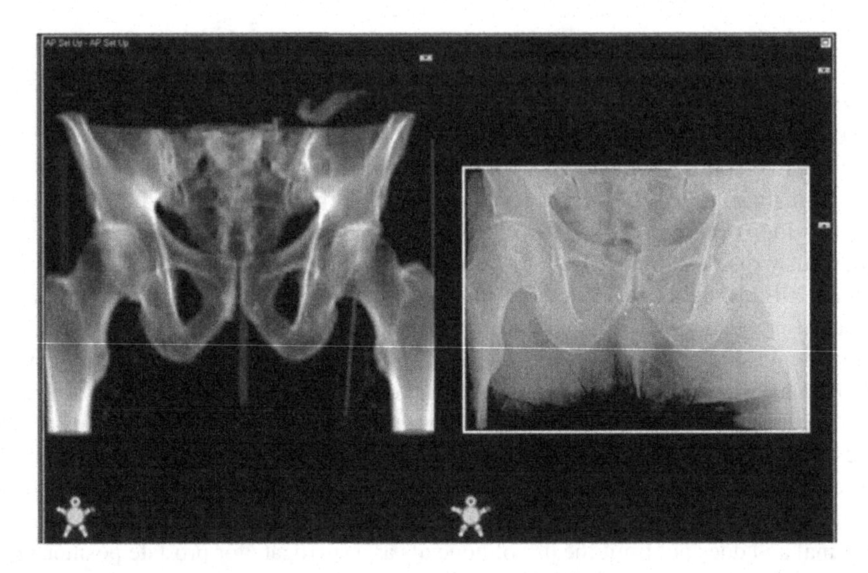

FIGURE 3.1 Fiducial markers used for verification. The left-hand image shows the planning digitally reproduced radiograph (DRR), and the right-hand image shows three fiducial markers clearly visible. Image courtesy of Varian Medical Systems.

Due to the simplicity of the image match when using fiducial markers, there has been found to be lower interoperator variability with match results (Moseley et al. 2007; O'Neil et al. 2016). Interoperator variability is any variability or difference seen in imaging results between two equally trained and competent treating staff members. This variability can lead to a lowered confidence in the image match result and extends the time the patient is on the treatment couch while an agreement is reached, or a further opinion is sought.

If fiducial marker match results show the prostate position to be outside of the departmental agreed tolerance, there may be a need for further imaging to be completed. As the imaging field size and modality when using fiducials does not allow for soft tissue visibility, there is no ability to assess reasons for such placement discrepancy. This may lead to a need to use volumetric imaging for that particular treatment. Before any large corrections can be applied to the treatment position, the reason for the discrepancy should be investigated. Certain positional errors cannot be rectified using standard treatment couch shifts or prostate rotations for example (Rijkhorstx et al. 2007), and therefore the cause of the displacement needs to be rectified, for example, bladder or bowel state. The need to use volumetric imaging following a fiducial match with planar images will increase both the time the patient is in the treatment position and the dose they receive. The likelihood of this occurring will need to be considered in contrast to using volumetric imaging as the main imaging modality.

In using fiducial markers and planar imaging as the main modality, the imaging process for the patient can be quicker than volumetric imaging (Eren et al. 2020); the dose received is lower as only two planar images are taken (Deegan et al. 2015) and image matching is a simpler task, reducing the interoperator variability (Schiffner et al. 2007).

3.3 VOLUMETRIC IMAGING CONE-BEAM COMPUTED TOMOGRAPHY (CBCT)

The use of volumetric imaging does not require the use of any surrogates as the prostate itself will be visible and can be used for image matching. Image-guided radiotherapy is recommended for prostate treatments (RCR 2021), and the visualisation of the prostate as well as the surrounding tissues and organs in the pelvis has advantages. (Moseley et al. 2007; Eren et al. 2020). Alongside direct visualisation of the prostate, other structures included in the target volume are visible. These may include the seminal vesicles and surrounding lymph nodes. Also visible are the bladder and bowel. Any discrepancy in the state or size of these organs can be assessed, and the effect any changes have on the prostate position can be addressed. Should a change in bladder state alter the prostate position, it can be assessed directly and rectified. Changes in prostate position caused by bladder or bowel state cannot always be corrected using treatment couch shifts. Rotations, for example, may not be rectified without a six degrees of freedom couch top and, even then, there is a limit to the rotation that can be applied (Rijkhorstx et al. 2007). Therefore, the cause of the position change should be rectified directly by repositioning the patient or by correcting the bladder or bowel state. Being able to visualise changes to bladder and bowel volume will allow for this assessment and correction to be completed (Moseley et al. 2007; O'Neil et al. 2016). A main consideration for this however relates to the additional dose delivered when using CBCT (Ariyaratne et al. 2016; Deegan et al. 2015). Should the bladder or bowel state be shown to be inconsistent with the planned state, the patient may be required to leave the treatment room to address this. Should this happen, when the patient is repositioned on the treatment couch, imaging will need to be repeated (Ariyaratne et al. 2016). This will result in two CBCTs taken for this treatment and, although justified, the increased dose must be considered. One method to reduce the dose given in this situation is the addition of other imaging modalities to assess bladder and bowel state prior to the CBCT, such as ultrasound.

The added dose received using CBCT when compared with planar imaging and fiducial markers may be offset by a reduction in the treatment margins and therefore the dose received by the rectum (Ariyaratne et al. 2016; Maund et al. 2014). However, margin reduction has also been suggested when using fiducial marker matches for daily prostate verification (Skarsgard et al. 2010; Deegan et al. 2015).

Interoperator variability in matching and image analysis results can be an issue when using soft tissue image registration with volumetric imaging (Deegan et al. 2015; Moseley et al. 2007). This issue may also limit the ability to reduce margins when using volumetric imaging (Deegan et al. 2015). When compared with soft tissue and the use of CBCT, fiducial marker matches have been shown to reduce interoperator variability due to the simplicity of matching to markers instead of anatomical structures (Moseley et al. 2007).

Time efficiency of the imaging method has been discussed in relation to fiducial markers. CBCT acquisition and analysis can add to the time the patient is in the treatment room; this may add an extra burden to the department in terms of patient throughput (Eren et al. 2020). However, the greatest concern for time taken is the patient's stability in the treatment position. The longer the treatment time, the higher the probability of any movement occurring (Vanhanen et al. 2020), either intentional, by the patient due to discomfort, or unintentional movement, due to internal organ motion of the prostate itself, mainly caused by a change in bladder or bowel position (Steiner et al. 2013; Boda-Heggemenn et al. 2008). A further time concern is the internal immobilisation used for the patient; patients treated with a full bladder may struggle to hold this bladder state for the treatment duration, and therefore any additional time taken to verify and deliver treatment may make this state unachievable, especially as the patient progresses through treatment and experiences bladder side effects. Gurjar et al. suggested an increased need for volumetric imaging for patients unable to hold a full bladder; this is needed to be able to verify the effect of a different bladder size on the position of the prostate (Gurjar et al. 2020).

Standard on-treatment imaging is acquired and assessed directly prior to treatment delivery. This online method of image evaluation is considered the standard due to its ability to apply positional corrections before treatment commences, allowing the accuracy of treatment delivery to be established (RCR 2021). If using a daily imaging schedule, both random and systematic errors can be evaluated and corrected before any error in treatment delivery occurs (RCR 2021). Although online image verification can reduce positional errors in treatment, care must be taken when using these on-treatment images to inform a reduction in treatment margins. The prostate is known to be a moveable structure and can experience positional shifts during treatment delivery (Chasseray et al. 2020; Boda-Heggemenn et al. 2008); this issue can be exacerbated if a longer treatment time is

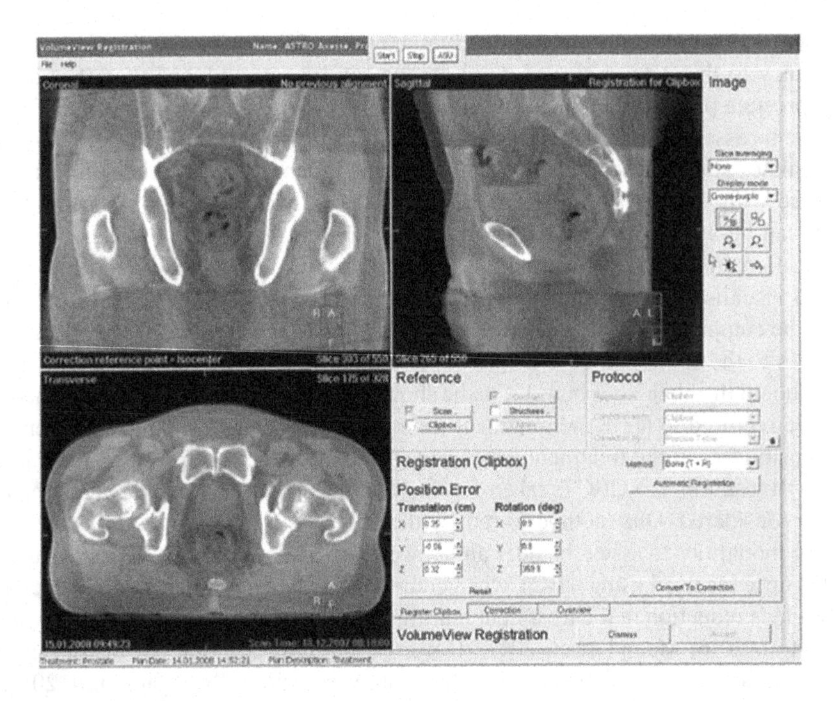

FIGURE 3.2 CBCT showing soft tissue match. Image courtesy of Elekta.

required, for example if using step-and-shoot intensity modulated radiotherapy (IMRT) or stereotactic body radiotherapy (SBRT) treatments (Chasseray et al. 2020; Vanhanen et al. 2020). In these circumstances, consideration should be given to repeating the images during treatment delivery, motion tracking or increased staff training to ensure images are evaluated quickly or there will be the possibility that margins cannot be reduced due to the need to account for intrafractional shift of anatomy (Ariyaratne et al. 2016; Skarsgard et al. 2010).

3.4 ULTRASOUND

The most relevant advantage of ultrasound (US) use is the lack of ionising radiation. US images do not need to be justified in the same way that radiation images do, and there are no limits applied to the frequency or number of images acquired. The use of US does require the purchase of additional equipment in the treatment room and involves extra training for radiographers. Despite these extra considerations, US guidance for use in radiotherapy has been produced (RCR 2021). US for prostate treatments can be considered as two separate processes: its use to visualise and confirm the location of the prostate itself and the ability to confirm bladder size and state. This section will focus on using US to confirm only bladder state.

CBCT and planar imaging used to verify prostate position both involve the use of ionising radiation and therefore the patient will receive an additional radiation dose associated with the imaging modality. When a discrepancy is seen in prostate position, it is generally due to inconsistencies in bladder or bowel position affecting the position or deforming the shape of the prostate. This issue will need to be rectified and the patient re-imaged, which will result in an extra radiation dose delivered to the patient when using CBCT or planar imaging. However, if bladder state and size can be confirmed prior to radiation being delivered, this added radiation dose will be avoided (Reilly et al. 2020). The bladder state can be confirmed as being consistent with the planning state with the use of US, prior to any radiation being delivered to the patient. This would then allow for the CBCT or planar imaging to focus on the prostate position without the effect of bladder inconsistencies (Cramp et al. 2016).

US imaging involves the use of a probe placed firmly onto the patient's skin; this can be completed using a transabdominal ultrasound (TAUS) method. For TAUS, the US probe is placed against the patient's abdomen and the procedure is described as being non-invasive as the probe is kept to the patient's surface. US can also be conducted using a transperineal (TPUS) or transrectal (TRUS) approach. These methods will not require the sound beam to pass through the bladder and are generally used for prostate visualisation; therefore, they will not be included in the discussion here.

The advantage of US not involving radiation must be considered alongside the potential limitations to US use. High interoperator and intraoperator variability have both been discussed as a limitation to US use (Van der Meer et al. 2013; Fontanarosa et al. 2015; Johnston et al. 2008). Inter- and intraoperator variability describes a difference seen between two operators performing the same task (interoperator variability) and between results when one operator performs the same task twice (intraoperator variability). These differences in results can lead to a lack of confidence in the use of US as a verification method; however, increased staff training has shown to reduce this variability (Johnston et al. 2008; Robinson et al. 2012). The increased training burden placed on a radiotherapy department coupled with the increased cost of the equipment when considering the use of US may prove to be too high to implement a US protocol for prostate patients.

The length of time the patient is on the treatment bed should be kept as short as possible due to any intrafractional movement of the prostate (RCR 2021). Implementing a treatment protocol that adds an additional stage to the treatment setup and verification will add to the time the patient is in the treatment position. This can be especially crucial for those patients treated with a full bladder. The US scan for bladder consistency can be completed either in the treatment room in the position for treatment or it may be done in a separate clinic room prior to patients being placed in position for treatment. The latter may prove to be more time efficient for the treating staff as LINAC in-room time is not extended. However, the patient will be required to hold a full bladder state for the US scan and throughout the treatment. If the US is completed in a separate room, problems may arise if there is then a delay in accessing the treatment room. Any delay requiring the patient to need to void and refill their bladder will represent only a time inconvenience as there is no radiation dose associated with US imaging, so that there are no limitations to the number of repeat scans carried out.

US images are acquired using a probe placed externally on the patient's skin, on the abdomen in this case. Sound waves used to create the image travel differently through air than via soft tissue. The presence of air in the image can cause difficulties in visualising internal structures located behind areas where air is present. Due to this the probe must be in close contact with the patient's skin and increasing the pressure placed on the probe will minimise any air trapped between it and the patient. Increasing the probe pressure, however, can lead to distortion of the underlying anatomy (Artignan et al. 2004; Camps et al. 2018). The issue of probe pressure is particularly relevant to bladder state and prostate position, as most patients are treated with a full bladder. In addition to the pressure of the probe on the abdomen potentially leading to increased discomfort for the patient, probe pressure can distort the shape of the bladder making it difficult to confirm the bladder state when compared with images taken at planning (Camps et al. 2018). Staff training will be required to ensure that sufficient probe pressure is used but anatomy distortion is minimal.

Despite the limitations and concerns relating to the use of US for bladder verification, US has been reported as being well tolerated by prostate patients (Brown et al. 2020). Compared with the use of fiducial markers or CBCT, US for bladder verification is non-invasive and requires no radiation dose to be delivered.

When considering the imaging method and modality to be used, they should be discussed in relation to other treatment equipment available and protocols to be used. Imaging methods cannot be successfully implemented without consideration of the immobilisation used and the margins applied at the planning stage. These will impact the imaging requirements and, in turn, the imaging used will affect the margins and immobilisation required.

REFERENCES

Ariyaratne, H., Chesham, H., Pettingell, J., Alonzi, R. 2016. Image-guided radiotherapy for prostate cancer with cone beam CT: Dosimetric effects of imaging frequency and PTV margin. *Radiotherapy and Oncology*. 121: 103–108.

Artignan, X., Smitsmans, M. H. P., Lebesque, J. V., et al. 2004. Online ultrasound image guidance for radiotherapy of prostate cancer: Impact of image acquisition on prostate displacement. *International Journal of Radiation Oncology, Biology, Physics*. 59(2): 595–601.

Balter, J. M., Sandler, H. M., Lam, K., et al. 1995. Measurement of prostate movement over the course of routine radiotherapy using implanted markers. *International Journal of Radiation Oncology, Biology, Physics*. 31(1): 113–118.

Boda-Heggemenn, J., Kohler, F. M., Wertz, H., et al. 2008. Intrafraction motion of the prostate during an IMRT session: A fiducial-based 3D measurement with cone-beam CT. *Radiation Oncology*. 3: 37.

Brown, A., Pain, T., Preston, R. 2020. Patient perceptions and preferences about fiducial markers and ultrasound motion monitoring procedures in radiation therapy treatment. *Journal of Medical Radiation Sciences*. 1–7.

Camps, S. M., Fontanarosa, D., de With, P. H., et al. 2018. The use of ultrasound imaging in the external beam radiotherapy workflow of prostate cancer patients. *Biomed Research International*. 2018: 7569590.

Chan, M. F., Cohen, G. N., Deasy, J. O. 2015. Qualitative evaluation of fiducial markers for radiotherapy imaging. *Technology in Cancer Research and Treatment*. 14(3): 298–304.

Chasseray, M., Dissaux, G., Lucia, F., et al. 2020. Kilovoltage intrafraction monitoring during normofractionated prostate cancer radiotherapy. *Cancer/Radiotherapie*. 24: 99–105.

Cramp, L., Connors, V., Wood, M., et al. 2016. Use of a prospective cohort study in the development of a bladder scanning protocol to assist in bladder filling consistency for prostate cancer patients receiving radiation therapy. *Journal of Medical Radiation Sciences*. 63: 179–185.

Crook, J. M., Raymond, Y., Salhani, D., et al. 1995. Prostate motion during standard radiotherapy as assessed by fiducial markers. *Radiotherapy and Oncology*. 37: 35–42.

Das, S., Liu, T., Jani, A. B., et al. 2014, December. Comparison of image-guided radiotherapy technologies for prostate cancer. *American Journal of Clinical Oncology*. 37(6): 616–623.

Deegan, T., Owen, R., Holt, T., et al. 2015 Assessment of cone beam CT registration for prostate radiation therapy: Fiducial marker and soft tissue methods. *Journal of Medical Imaging and Radiation Oncology*. 59: 91–98.

Delouya, G., Carrier, J-F., Beliveau-Nadeau, D., et al. 2010. Migration of fiducial markers and its influence on the matching quality in external beam radiation for prostate cancer. *Radiotherapy and Oncology*. 96: 43–47.

Eren, M. F., Oksuz, D. C., Sayan, M., et al. 2020. Comparison of kV radiographs and kV-Cone-beam computed tomography image-guided radiotherapy methods with and without implanted fiducials in prostate cancer. *Cureus*. 12(8)

Fontanarosa, D., Van der Meer, S., Bamber, J. et al. 2015. Review of ultrasound image guidance in external beam radiotherapy: I. Treatment planning and inter-fraction motion management. *Physics in Medicine and Biology*. 60: R77–R114.

Ghaffari, H., Navaser, M., Mofid, B., et al. 2019, March 11. Fiducial markers in prostate cancer image-guided radiotherapy. *Medical Journal of Islamic Republic of Iran*. 33: 15.

Gurjar, O. P., Arya, R., Goyal, H. 2020. A study on prostate movement and dosimetric variation because of bladder and rectum volume changes during the course of image-guided radiotherapy in prostate cancer. *Prostate International*. 8: 91–97.

Johnston, H., Hilts, M., Beckham, W., Berthelet, E. 2008, June. 3D ultrasound for prostate localisation in radiation therapy: A comparison with implanted markers. *Medical Physics*. 35(6).

King, B. L., Butler, W. M., Merrick, G. S., et al. 2011. Electromagnetic transponders indicate prostate size increase followed by decrease during the course of external beam radiation therapy. *International Journal of Radiation Oncology, Biology, Physics*. 79(5): 1350–1357.

Kupelian, P. A., Willoughby, T. R., Sanford, M. S., et al. 2005. Intraprostatic fiducials for localisation of the prostate gland: Monitoring intermarker distances during radiation therapy to test for marker stability. *International Journal of Radiation Oncology, Biology, Physics*. 62(5): 1291–1296.

Loh, J., Baker, K., Sridharan, S., et al. 2015. Infections after fiducial marker implantation for prostate radiotherapy: Are we underestimating the risk? *Radiation Oncology*. 10: 38.

Maund, I. F., Benson, R. J., Fairfoul, J., et al. 2014. Image-guided radiotherapy of the prostate using daily CBCT: The feasibility and likely benefit of implementing a margin reduction. *British Journal of Radiology*. 87: 20140459.

McNair, H., Hansen, V. N., Parker, C. C., et al. 2008. A comparison of the use of bony anatomy and internal markers for offline verification and an evaluation of the potential benefit of online and offline verification protocols for prostate radiotherapy. *International Journal of Radiation Oncology, Biology, Physics*. 71(1): 41–50.

Moseley, D. J., White, E. A., Wiltshire, K. L., et al. 2007. Comparison of localisation performance with implanted fiducial markers and cone-beam computed tomography for on-line image-guided radiotherapy of the prostate. *International Journal of Radiation Oncology, Biology, Physics*. 67(3): 942–953.

Munoz, F., Fiandra, C., Franco, P., et al. 2012. Tracking target position variability using intraprostatic fiducial markers and electronic portal imaging in prostate cancer radiotherapy. *Radiologia Medica*. 117: 1057–1070.

O'Neil, A. G. M., Jain, S., Hounsell, A. R., O'Sullivan, J. M., 2016. Fiducial marker guided prostate radiotherapy: A review. *British Journal of Radiology*. 89(20160296)

RCR (Royal College of Radiologists). 2021. *On Target 2: Updated Guidance For Image-Guided Radiotherapy*. London, UK: RCR.

Reilly, M., Ariani, R., Thio, E., et al. 2020. Daily ultrasound imaging for patients undergoing prostatectomy radiation therapy predicts and ensures dosimetric endpoints. *Advances In Radiation Oncology*. 5: 1206–1212.

Rijkhorstx, E-J., van Herk, M., Lebesque, J. V., Sonke, J-J. 2007. Strategy for online correction of rotational organ motion for intensity-modulated radiotherapy of prostate cancer. *International Journal of Radiation Oncology, Biology, Physics*. 69(5): 1608–1617.

Robinson, D., Liu, D., Steciw, S., et al. 2012. An evaluation of the Clarity 3D ultrasound system for prostate localisation. *Journal of Applied Clinical Medical Physics*. 13(4): 100–112.

Schiffner, D. C., Gottschalke, A. R., Lometti, M., et al. 2007. Daily electronic portal imaging of implanted gold seed fiducials in patients undergoing radiotherapy after radical prostatectomy. *International Journal of Radiation Oncology, Biology, Physics*. 67 (2): 610–619.

Skarsgard, D., Cadman, P., El-Gayed, A., et al. 2010. Planning target volume margins for prostate radiotherapy using daily electronic portal imaging and implanted fiducial markers. *Radiation Oncology*. 5: 52.

Steiner, E., Georg, D., Goldner, G., Stock, M. 2013. Prostate and patient intrafraction motion: Impact on treatment time-dependant planning margins for patients with endorectal balloons. *International Journal of Radiation Oncology, Biology, Physics*. 86(4): 755–761.

Vanhanen, A., Poulson, P., Kapanen, M. 2020. Dosimetric effect of intrafraction motion and different localisation strategies in prostate SBRT. *Physica Medica*. 75: 58–68.

Van der Heide, U. A., Kotte, A. N. T. J., Dehnad H., et al. 2007. Analysis of fiducial marker-based position verification in the external beam radiotherapy of patients with prostate cancer. *Radiotherapy and Oncology*. 82: 38–45.

Van der Meer, S., Bloemen-van Gurp, E., Hermans, J., et al. 2013, July. Critical assessment of intramodality 3D ultrasound imaging for prostate IGRT compared to fiducial markers. *Medical Physics*. 40(7): 071707-1–071707-11.

Van der Wielen, G. J., Mutanga, T. F., Incrocci, L., et al. 2008. Deformation of prostate and seminal vesicles relative to intraprostatic fiducial markers. *International Journal of Radiation Oncology, Biology, Physics*. 72(5): 1604–1611.

Webster, A., Appelt, A. L., Eminowicz, G. 2020. Image-guided radiotherapy for pelvic cancers: A review of current evidence and clinical utilisation. *Clinical Oncology*. 32: 805–816.

4 IGRT Using Non-Ionising Radiation

4.1 INTRODUCTION

The basic use of non-ionising radiation methods for image-guided radiotherapy (IGRT) is covered in chapters 7 and 8 of Kirby and Calder (2019). Readers are directed to these chapters as a primer for initial study on these methods and technologies sections. Specifically sections 7.7 and 8.8 detail the technical and clinical practice principles, respectively, for non-ionising radiation methods such as the use of ultrasound (US), surface guidance and electromagnetic (EM) transponders for IGRT. Excellent information is also given within the current UK national guidance, On-Target 2 (RCR 2021), within chapter 4.

The main advantage here over X-ray based methods of on-treatment IGRT is that the patient does not receive any increase in radiation dose through imaging. These methods of imaging can therefore be used frequently, with no dose justification required, often for monitoring/tracking intrafractional motion during treatment delivery.

Within this chapter, we shall examine the use of US, EM transponders and surface-guided radiotherapy (SGRT) methods. Magnetic Resonance Imaging (MRI) is addressed in Chapter 5.

4.2 ULTRASOUND

In-room ultrasound (US) has been in use in radiotherapy since the early 1990's, beginning almost exclusively with use during the treatment of prostate cancer (PCa) patients. The greatest volume of evidence for IGRT purposes is for PCa. Technicalities are covered in Kirby and Calder (2019 chapters 7 and 8), and in chapter 4 of RCR (2021). A key recent review paper of its use for PCa patients is published by Camps et al. (2018). In brief, contours delineated from computed tomography (CT) simulation and pre-treatment computer planning are manually or automatically overlaid onto real-time US images – through an operator in the treatment room (during patient set-up) or through specialised mounts for the ultrasound probe attached to the treatment couch, so that ultrasound images can be acquired continuously during treatment delivery for continuous motion monitoring intrafractionally, without increasing the concomitant dose burden. The Clarity system is just such a method. With both methods, target and organ at risk (OAR) volumes can be compared with planned volumes and set-up adjustments/decisions to treat being made prior to treatment delivery. In the case of the Clarity system for continuous prostate position monitoring, treatment can be interrupted if movement is detected above a certain threshold for a certain amount of time (De Los Santos et al. 2013; Camps et al. 2018; Bertholet et al. 2019; Kirby and Calder 2019; RCR 2021).

There are three main methods of acquiring ultrasound (US) data from the prostate: (i) TAUS – Transabdominal US; (ii) TRUS – Transrectal US; and (iii) TPUS – Transperineal US. The two main methods used for IGRT are TAUS and TPUS; both being non-invasive and viable for continuous monitoring (although it is mainly TPUS which gives the best results (De Los Santos et al., 2013; Camps et al. 2018; Ghadjar et al. 2019).

TAUS has the ability to measure prostate volumes by making use of the acoustic window presented by a filled bladder (which also becomes a check of bladder status, since it needs to be reasonably full to prevent acoustic impedance of the ultrasound waves). Use is made of suprapubic transducers to obtain 2D or 3D US images of the prostate target volume while the patient is in a treatment position, then fused with the computed tomography (CT) planning scans and/or matched

DOI: 10.1201/9781003050988-4

with contours derived from the CT-based treatment plan, and set-up corrections can be performed on-treatment with reference to the treatment isocentre. Most procedures take about 5–7 min, for US image acquisition, analysis and set-up correction (De Los Santos et al. 2013; Camps et al. 2018).

For TAUS, being "far" from the prostate, image quality can be affected, an issue made more challenging by larger/obese patients. Ideal for imaging at the time of set-up and therefore interfractional imaging, probe positioning makes it more challenging for intrafractional motion monitoring during treatment delivery by potentially being in beam paths, especially for volumetric modulated arc therapy (VMAT) techniques (Camps et al. 2018; Ghadjar et al. 2019). The more advanced systems involve the generation of US-based deformable registration to create daily pseudo-CT images for verification (Camps et al. 2018).

TPUS is also non-invasive and involves a probe position in direct contact with the perineum. It also enables a measurement of prostate volume and positioning but without the need for an acoustic window through the bladder. With the smaller distance involved to the prostate, image quality can be good, and, with its positioning outside the main beam path, intrafractional monitoring is possible. Figure 4.1 shows the Clarity TPUS system used for prostate on-treatment image guidance, for set up and continuous intrafractional monitoring. The systems can be used with automated couch correction or gating (via an action threshold, e.g., 3 mm beyond tolerance for 10 seconds). Using an internal swept transducer array, 3D imaging is possible without moving the probe or needing user intervention. With automatic template matching during continuous monitoring, the prostate position can be estimated with respect to planning references. The probe is matched to the room coordinates (and therefore the isocentre) by way of optical infrared (IR) markers on the probe, which are imaged by in-room stereoscopic IR cameras at a rate of about 0.5 Hz typically (Camps et al. 2018; Bertholet et al. 2019).

TPUS studies show a greater correlation and therefore more confidence in regarding interchangeability compared to TAUS, but neither could be considered equivalent to other more established and frequently used X-ray based IGRT methods. Limits of agreement can vary by as much as 3–9 mm; and probe pressure can be an issue during TPUS (see below) (Dang et al. 2018; Ghadjar et al. 2019). Because of the real-time monitoring capabilities presented, research has been conducted

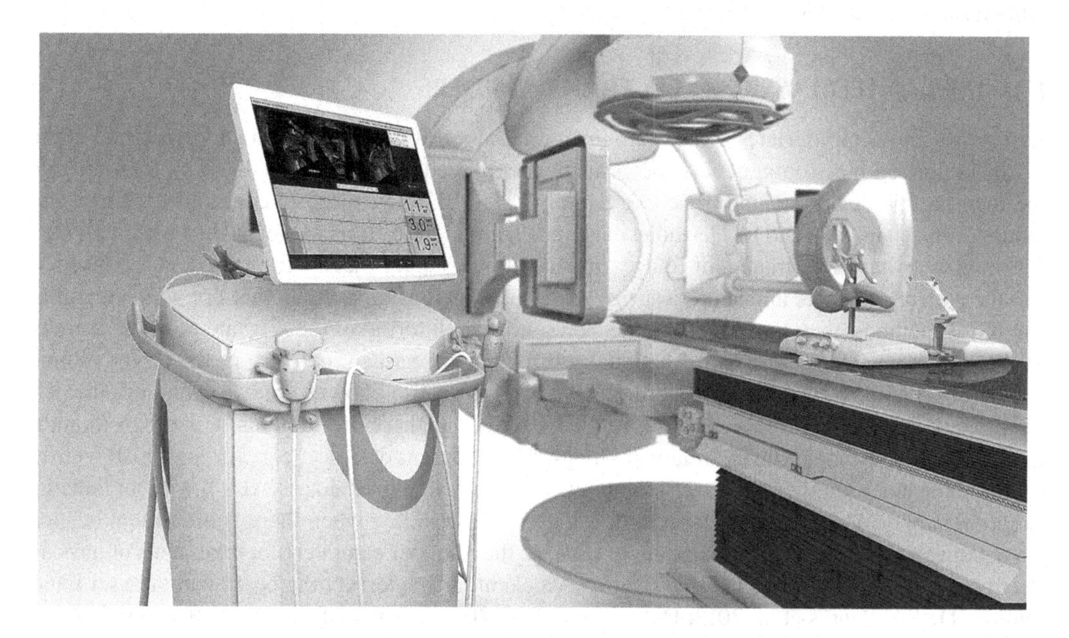

FIGURE 4.1 The Clarity TPUS ultrasound system used for trans-perineal on-treatment image guidance to set up geometric verification of the prostate and also continuous intrafractional monitoring of prostate position. Image courtesy of Elekta.

into linking motion detection with automatic multi-leaf collimator (MLC) leaf tracking, thereby enabling some form of treatment adaptation in real time to follow the relatively slow intrafractional motion associated with the prostate (Bertholet et al. 2019).

The potential problem of probe pressure has been mentioned above and was best investigated by Li et al. (2017) for TPUS. Tissue deformation and relaxation have been noticed both after and during US probe application in TPUS, at the time of planning and during treatment delivery. The result is to cause the target and at-risk organs to be displaced and drift during treatment delivery. Results of experiments show that from the initial contact of the probe with the skin surface of the perineum, during use, a further displacement of 5–10 mm is normally needed to achieve good acoustic contact and thereby high-quality US images. When this is done, it can cause a shift in the prostate of about 2–4 mm in the cranial direction. Both tissue compressions and prostate displacements were clearly visible in the study of Li (2017). For normal clinical use, therefore, slight probe pressure is recommended to improve the image quality, but excessive pressure should be avoided otherwise it can cause displacements of the prostate and OARs (Li et al. 2017).

Comparisons have been made between non-ionising radiation methods like US and surface guided radiotherapy (SGRT) (AlignRT). The results of the Krengli et al. (2016) study, comparing Clarity (using TAUS) and surface guidance, are shared in Section 4.4 below.

In more recent work, Richter et al. (2020) studied intrafractional motion monitoring during VMAT treatments using over 400 US-monitored fractions using Clarity. Data were analysed retrospectively for the 14 patients involved, assessing three directions of displacement (superior/inferior (SI), left/right (LR), anterior/posterior (AP)). Planning margins for the patients were 5–10 mm for the 33 fraction treatments delivered using VMAT. CBCT was used as the primary on-treatment image guidance for set-up (fiducial markers (FMs) not used in this study). Results showed mean deviations all to be less than 1 mm, although some large deviations were recorded (8 mm) but only very infrequently (less than 0.1% of scans). Movement was within +/- 2 mm in 99%, 98.1%, and 96.6% of the treatment time in SI, LR and AP directions, respectively. For monitoring scans of 100 seconds, median displacements increased from 0.2 to 0.8 mm, with the maximum displacement changing from 5.2 to 7.8 mm. Most displacements (99%) were deemed to be within the accepted, applied safety margin for the prostate treatment.

4.3 IMPLANTED/INSERTED EM TRANSPONDERS

Perhaps the most appropriate of the non-ionising radiation IGRT methods for PCa patients is the use of implanted electromagnetic (EM) transponders. Technical details of the way they operate and their function can be found in de Los Santos et al. (2013), D'Ambrosio et al. (2012), Kirby and Calder (2019, chapters 7 and 8), Bertholet et al. (2019) and RCR (2021, chapter 4). Just like implanted FMs for X-ray-based on-treatment image guidance (see Chapter 3), EM transponders are inserted into the patient's prostate and act as a surrogate for prostate position. Like other non-ionising radiation IGRT methods, they provide real-time, continuous monitoring of the prostate position, thereby producing set-up information intrafractionally throughout treatment delivery of the target volume (D'Ambrosio et al. 2012; Bertholet et al. 2019). The prostate is known to move (non-periodically) during set-up and treatment delivery (see Chapter 5), and therefore systems which can monitor and track that movement during relatively long-term treatment delivery (e.g., through SBRT ultra-hypofractionated radiotherapy (UHF) treatments or adaptive radiotherapy (ART)) without ionising radiation may be advantageous, especially if they are able to flag when there is movement beyond pre-set thresholds (over a certain amount of time) so that the user can interrupt treatments or have an electronic link directly to the LINAC (Linear Accelerator) to interrupt the beam without user intervention (De Los Santos et al. 2013; D'Ambrosio et al. 2012; Kirby and Calder 2019, chapters 7 and 9; Bertholet et al. 2019; RCR 2021). Information acquired intrafractionally can be used to inform the margins used in planning, to help ensure the clinical target volume (CTV) is always covered by the high-dose volume (Chaurasia et al. 2018).

Two main commercial systems are in use: the Calypso system and the RayPilot system. Here we will consider first the Calypso system.

4.3.1 The Calypso EM Transponder system

Although full technical details can be found in the references above, briefly, the Calypso system uses a mobile emitter/receiver array positioned above the patient during treatment delivery. The position of the array can be tracked in the treatment room (and therefore linked to the isocentre position of the LINAC), using ceiling-mounted IR camera systems together with associated computer hardware and software. Figure 4.2 shows the Calypso detector array in position above the patient for prostate initial set-up and continuous intrafractional position monitoring throughout treatment delivery.

Transponders are implanted into the prostate before CT simulation so that their positions can then be determined with respect to the isocentre on the patient's treatment plan. When calibrated positionally within the treatment room, then their position can be accurately pinpointed (within sub-millimetre accuracy) when the patient is on the couch and throughout set-up (which is aided by using the Calypso electromagnetic (EM) transponder system) and then throughout treatment intrafractionally (De Los Santos et al. 2013; D'Ambrosio et al. 2012; Willoughby et al. 2006; Bertholet et al. 2019).

The Calypso system is the most mature of the two EM transponder systems, and therefore there are many publications about its commissioning and clinical utility, with other studies often using the intrafractional data acquired with Calypso as a gold standard reference for intrafractional movement modelling, ART studies and margin calculations (see Chapter 5).

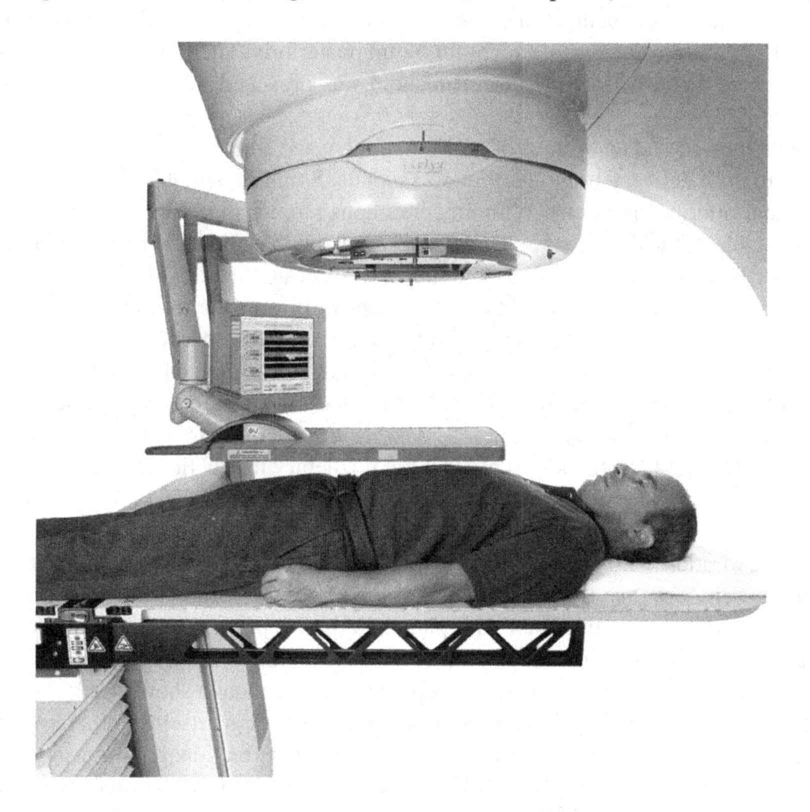

FIGURE 4.2 A photograph of the Varian Calypso on-treatment verification system, which uses inplanted beacon trasnponders for geometric verification on-treatment. The detector array is suspended above and close to the patient for detection of the trasnponder's position; for intiial patient set-up, and then for continuous intrafractional position monitoring throughout treatment delivery. Image courtesy of Varian Medical Systems.

The continuous monitoring capability (ten times per second) of the system (without using ionising radiation) has enabled systematic investigations of prostate motion patterns, with some studies showing strong cranial and anterior motion. There is an increased likelihood of small (3–5 mm) displacements with treatment time, and less likely large displacements (7–10 mm), together with some respiration-induced prostate motion for prone patients compared with supine patients (Dang et al. 2018; Bertholet et al. 2019). Studies have also shown the feasibility of margin reduction for conventional prostate treatments from typically 10 mm (without monitoring and correction) down to 2–3 mm (Dang et al. 2018). It also makes possible some unique approaches, e.g., allowing for prostate tracking and potentially adapting treatments using multi-leaf collimator (MLC) shape adaptation in real time (Colvill et al. 2016; Ghadjar et al. 2019).

There are a number of drawbacks to the Calypso system. The system requires a specialised couch top; there can be a lack of flexibility in moving between treatment rooms; there is a limited transponder detection volume (closeness of the detection array can be an issue for some treatment techniques); patients with hip prostheses or large metal implants and patients with pacemakers or other implanted EM devices can be excluded from treatment; larger patients (which would entail the transponders being outside the tracking range (> 27 cm) from the array) could also be excluded; and artefacts can be present during magnetic resonance (MR) imaging caused by the transponders (De Los Santos et al. 2013; D'Ambrosio et al. 2012; Dang et al. 2018; Bertholet et al. 2019).

Some of the earliest studies made use of the continuous monitoring aspects for SBRT prostate patients using 2 mm thresholds (Lovelock et al. 2015) in the Calypso system. In an study involving 89 patients, 32.5–40 Gy was delivered in five treatment fractions. Patients were treated with a full bladder, but no bowel instructions. Three EM transponders were implanted one week prior to CT simulation. Typical treatment arrangement involved a nine field IMRT delivery. Patients were planned with PTV margins of 5 mm, except for 3 mm posteriorly. Therapists' interruptions of treatment delivery were made when the 2-mm threshold was breached, if necessary with subsequent couch correction.

Their results showed that an average of 1.7 interventions per fraction was needed, with a concomitant increase in dose delivery time of about 65 seconds. Small systematic drifts were observed in the posterior and inferior directions. Without continuous monitoring, about 10% of patients would not have acceptable clinical coverage. As a function of time (without correction), margins would need to increase by 2 mm every 5 mins, from the start of set-up imaging. Reductions in set-up time and more rapid delivery methods would likely further minimise the intrafraction motion experienced and target drifts observed, further reducing the target under dosage.

In comparing Calypso with other methods (e.g., cone-beam computed tomography (CBCT) and kilovoltage (kV) imaging), Foster et al. (2012) examined all three image guidance technologies. Prostate patients had gold fiducial markers (FMs), 6 mm margins (4 mm posteriorly) and were treated with 79.2 Gy in 44 fractions; full bladder but no dietary instructions. Patients were localised with lasers and skin marks; then set up with Calypso and either CBCT or kV orthogonal image pairs.

In patient localisation with both CBCT/kV imaging and the Calypso system, average localisation differences were less than 0.8 mm, deemed to be in excellent agreement. Greater uncertainty was found from CBCT positioning than with planar kV image pairs, though the difference was not deemed to be clinically significant and did not outweigh the clinical usefulness of the extra anatomical information from volumetric methods. The Calypso system, though showing excellent agreement with CBCT/kV imaging, did not provide information about the target itself or other organs at risk.

More recently, Gorovets et al. (2020) compared kV-MV intrafractional monitoring with Calypso (see Chapter 5). They used LINAC-based kV-MV monitoring for intrafractional motion in prostate stereotactic body radiotherapy (SBRT), ultra-hypofractionated radiotherapy (UHF) patients (40 Gy in five fractions) in a study with 193 patients. The initial set-up involved orthogonal kV image pairs and FMs and CBCT, with motion then tracked during VMAT (typically two full arcs). Treatment was interrupted if motion >1.5–2 mm was detected and the patient was repositioned.

Positioning data were compared with Calypso data from a previous study, with Calypso monitoring showing similar results (and therefore using Calypso as the reference for comparison). The LINAC-based kV-MV system was successfully implemented for intrafraction motion management, with prostate movement < 3 mm in any direction (median overall treatment time 8.2 mins (4.2–44.8 mins)). Motion > 3mm and > 5mm in any direction was observed in 32.3% and 10.2% of the fractions, respectively.

The possible use of ART has been explored by Olsen et al. (2012). They used the tracking data acquired and analysed for patients treated with step and shoot (SnS) intensity modulated radiotherapy (IMRT) and 5 mm planning target volume (PTV) margins using Calypso. Further plans were generated with 0 mm and 3 mm margins and then assessed when modelled with Calypso tracking data. It developed into an automated process to evaluate PTV margin adequacy for motion applied for the first three, five and ten fractions and for the entire treatment course. The model can be used to assess plans and margin suitability based on the evaluations using real-time tracking data from Calypso for prostate patients.

Vanhanen et al. (2020) shared experiences of the use of Calypso with prostate SBR UHF patients, treatment with five fraction (7 or 7.25 Gy per fraction) techniques every other day. Initial set-up was with skin marks and lasers; Calypso was used for initial localisation and subsequent motion tracking. CBCT was used to check bladder and rectum status and matched to planning CT slices using transponders as FMs; if displacement was detected between initial set-up and CBCT, displacement was corrected with couch shifts. After CBCT, an orthogonal kV pair was used to further confirm localisation using transponders as FMs. kV imaging was conducted purely for comparisons and was not deemed necessary if Calypso was used for set-up.

Treatment started and, if motion exceeded 2 mm posterior, 3 mm elsewhere, treatment was gated off automatically. Calypso adaptive couch repositioning was used if the prostate did not return to within the tolerances. Calypso data were then used to examine other possible treatment strategies without intrafraction motion monitoring/correction (Vanhanen et al. 2020).

Results showed that, due to intrafraction motion and time for CBCT acquisition and interpretation, CBCT-guided set-up without further correction may result in clinically relevant dose reductions in prostate SBRT for patients where the prostate motion was large. Additional position correction (with kV orthogonal imaging or Calypso) increases accuracy and is adequate for most fractions and patients. Continuous motion monitoring-based correction ensured that target dose coverage remained appropriate and minimised organ at risk (OAR) exposure for all fractions and patients (Vanhanen et al. 2020).

4.3.2 The RayPilot/HypoCath EM Transponder Systems

The second EM system to consider is that of RayPilot, which is now marketed in a different configuration and design as HypoCath, using a slightly different system of insertion and greater positioning stability.

The RayPilot system is slightly different, in that it is a wired implantable radiofrequency (RF) transmitter, receiving power through the wire from a couch top plate. The couch top is specialised, housing the receiving antennas which detect the transmitter position and orientation 30 times a second. The transmitter for the first system is implanted transperineally with implantation and explantation procedures considered safe and feasible, with the transmitter being removed after treatment (Braide et al. 2018; Trainer et al. 2020a, 2020b). The latest version (HypoCath) now makes use of a Foley catheter and balloon, inserted transurethrally. Earlier clinical studies with the transperineal version found position instabilities and recommended use combined with independent IGRT systems for daily localisation. HypoCath has no surgical intervention and helps localise the urethra as well as the prostate (Bertholet et al. 2019).

Initial clinical feasibility was investigated by Braide et al. (2018) with the first version (RayPilot), examining user and patient experiences and positional stability with respect to traditional FMs.

From an initial ten-patient trial (treated conventionally in 2 Gy fractions and set up with kV orthogonal imaging on the FMs), patients reported mild to moderate discomfort and impact on daily activities during implantation, but overall tolerance was good and so the procedure was generally considered feasible and safe. No real-time data were included in the study, but positioning was compared with the FMs. A variety of results were produced – with some 3D shifts 5–7 mm (4/10 patients) (mean > 2mm); three patients mean 3D shift < 2mm. Interfractional transmitter stability, though, was noted, causing a need for verification with further studies and used for real-time tracking only in combination with other daily set-up techniques (Braide et al. 2018)

Biston et al. (2019) compared RayPilot with the TPUS Clarity system for continuous monitoring in a phantom and patient study. The phantom study revealed no interference between the two systems and close positive correlation between the two methods. On-treatment, initial patient set up used CBCT and FM registration. Motion monitoring with RayPilot and Clarity was initiated during the CBCT scan. During treatment, a threshold of 3 mm was set, with irradiation halted if beyond this threshold for 15 seconds with both devices. If displacement lasted > 1 min, repositioning was undertaken using CBCT and FM registration; to minimise concomitant dose, a max of two intrafraction CBCTs were permitted in one fraction (Biston et al. 2019).

Strong correlation between the two methods was found in left/right (LR) direction with differences > 2 mm noted in less than 0.22% of the time and never exceeding five seconds; superior-inferior > 2 mm differences were monitored 6.5% of the time for three patients. Large prostate rotations, presence of bowel gas and EM transponder location in the prostate could explain these differences (Biston et al. 2019).

Vanhanen et al. (2018) compared the two EM transponder systems – Calypso and RayPilot – with kV FM-based imaging. Couch shifts from Calypso and RayPilot systems were compared with simultaneous FM-based orthogonal kV imaging for 582 fractions from 22 RayPilot patients and 335 fractions from 26 Calypso patients. Limits of agreement (from Bland-Altman analysis) were considerably greater for RayPilot and kV imaging compared with those between Calypso and kV imaging. Conclusions were that localisation with Calypso was comparable to kV imaging and the methods were deemed to be interchangeable; localisation accuracy with RayPilot was affected by transmitter instability (as noted by Braide et al. (2018)) and positioning of patients needs other image-guidance methods. RayPilot could be used for intrafractional motion monitoring, but initial set-up and positioning should be by other means (such as kV or CBCT imaging of FMs). The advantage of RayPilot is that it is removable after treatment, whereas the Calypso transponders could cause imaging artefacts in MR imaging used for follow-up.

Both Panizza et al. (2022) and Berchtold et al. (2023) have investigated the newer HypoCath system for dose-escalated SBRT UHF prostate patients. In contrast to the RayPilot system, HypoCath now makes use of a Foley catheter (and its balloon) for insertion and stable positioning via the patient's urethra. The catheter has within it coaxial wiring and a connector attaching to the detector plate. The blocking balloon within the bladder ensures safe and reproducible transmitter positioning. The catheter is inserted prior to planning CT and remains in place throughout the entire treatment (Berchtold et al. 2023).

For the work by Panizza et al. (2022), initial set-up was done using CBCT soft tissue matching, and motion was tracked throughout imaging and treatment. The EM transponder positioning was zeroed at the start of CBCT acquisition, and beam interruption was in place for motion beyond 2 mm; if > 15 seconds, then a new CBCT was acquired and the patient was re-set up (including a re-zero of HypoCath).

For the work by Berchtold et al. (2023), a 3 mm threshold was set for intrafractional motion monitoring and automatic radiation interruption; if beyond 3 mm for > 20 seconds, kV imaging was reinitiated and the HypoCath system was nullified.

Panizza et al. (2022) considered the EM tracking to be successfully implemented, with most displacements being < 2 mm in any direction; and a non-insignificant number of fractions with motion exceeding the defined threshold, which would have gone unnoticed (and uncorrected) without the intrafractional motion monitoring. For 45% of fractions, correction was mandated, but only in 18% was the beam interrupted. Total treatment time was on average 10.2 mins; with the prostate out of tolerance in 8% of the total session time, namely 4% in the set-up phase and 14% in the beam-on phase.

Berchtold et al. (2023) concluded that the system provided accurate and robust intrafractional motion monitoring, in conjunction with CBCT for initial patient set up; and could be used stand-alone for IGRT as an alternative to MR guided or FM real-time imaging.

4.4 SURFACE GUIDANCE

Like other non-ionising methods, surface-guided radiotherapy (SGRT) has the advantage of being an image guidance method for on-treatment geometric verification without the use of ionising radiation, and therefore lends itself to being used for continuous monitoring of the patient's surface (or part thereof) in real time for initial set up, on-treatment image guidance and then continuous monitoring for any intrafractional motion throughout the treatment delivery for a fraction, with no concomitant extra dose burden (D'Ambrosio et al. 2012; de Los Santos et al. 2013; RCR 2021; Wohlfahrt and Schellhammer, 2022). For an initial study into the technology and clinical applications of surface-guided methods, the reader is directed to sections 7.7 and 8.6 within Kirby and Calder (2019), and also to chapter 4 in RCR 2021.

Two products are perhaps the most mature in terms of SGRT within the UK, namely AlignRT (VisionRT, London, UK) and C-Rad Catalyst (C-RAD Positioning AB, Uppsala, Sweden). Both provide the ability to perform rapid surface imaging, using the principles of photogrammetry, generating and using for matching, 3-D surface models of the patient. In brief, two or more cameras (IR or near-visiblel) are mounted securely and rigidly in the room (CT simulator and/or treatment room) for the AlignRT system; for C-RAD, a line scanning approach is used. The reference (the desired treatment position) is derived from the CT simulation dataset, including the body contours and the surface relationship to the treatment room isocentre and where it is positioned within the target volume (De Los Santos et al. 2013; D'Ambrosio et al. 2012; Ghadjar et al. 2019; Kirby and Calder 2019; RCR 2021).

Rigid body transformation algorithms are used to match the model derived at the time of treatment setup and delivery with the reference model from CT simulation (although one can also use that acquired at the first fraction within the treatment room as the reference, post on-treatment image guidance using traditional x-ray methods). Computed displacements between the on-treatment and reference surface models may be translated into couch moves for correction of set-up, or used as flags to the user to interrupt treatment (when outside a pre-set tolerance), or interfaced directly to the Linac to interrupt radiation without user intervention (De Los Santos et al. 2013; Willoughby et al. 2006; D'Ambrosio et al. 2012; Wohlfahrt and Schellhammer, 2022).

Phantom studies and evaluation indicate that sub-millimetre precision is achievable. Clinically, image guidance for surface target volumes (such as breast treatments) shows the greatest utility and efficacy; its use for internal target volumes (such as the prostate) is less efficacious, but not without its use and merit; such as in aiding and making the initial set-up of the patient more speedy and efficient without traditional lasers and (often permanent) skin marks, as discussed in the following section. The systems can be fairly straightforward to implement, have a natural real-time capability and all without any radiation dose. Efforts to correlate patient's surface with internal target motion can be unreliable, especially for deep-seated lesions, making the ideal to combine SGRT with internal imaging for on-treatment image guidance (De Los Santos et al. 2013; D'Ambrosio et al. 2012; Ghadjar et al. 2019; RCR 2021; Kirby and Calder 2019; Wohlfahrt and Schellhammer 2022).

For the prostate, the tumour motion, intrafractionally, is non-periodic and less extreme, but perhaps more erratic compared with respiratory motion (Bertholet et al. 2019; Tudor et al. 2020; see Chapter 5). The feasibility and continuing impressive results from SBRT-style prostate treatments using hypofractionated (and ultra-hypofractionated) regimes, and also now adaptive radiotherapy

(ART) (see Chapters 5 and 8) depend upon the ability to maintain accurate localisation within reduced margins, throughout the treatment delivery (D'Ambrosio et al. 2012; de Los Santos et al. 2013). These regimes are therefore crucially dependent upon technologies that allow precise and real-time target localisation to continue to achieve reduced irradiated volumes of normal tissue and therefore maintain or improve still further the toxicity gains being evidenced from such techniques (see Chapter 9). Patients may well benefit from non-ionising continuous monitoring technologies and methods as a component of the overall image guidance used on-treatment (D'Ambrosio et al. 2012; Ghadjar et al. 2019; de Los Santos et al. 2013).

Within these efforts to find complementary IGRT solutions for prostate cancer (PCa) radiotherapy patients, SGRT methods are investigated and often compared against other more established on-treatment image guidance methods that can visualise internal anatomy or appropriate prostate surrogates. The next few papers outline some of the methods used and their results.

Pallotta et al. (2015) used a comparison of surface imaging, electronic portal (MV) imaging and skin mark set-up with respect to CBCT for pelvis patients. Using CBCT as the gold standard and the C-Rad system, 20 patients were investigated (ten pelvis, ten thorax). CT simulation was used for all patients with traditional tattoos (four, with three in transverse plane) after establishing the treatment isocentre. Initial patient alignment was with skin tattoos and lasers; position verification was performed for the first four fractions,acquiring an anterior/posterior, left/right pair of electronic portal images (EPIs), surface acquisition (SA), and a CBCT scan. Surface acquisition was performed during the EPI pair exposure. CBCT was used as the primary set-up method prior to treatment; registration results from EPIs (using a bony structure match) and SA were compared with CBCT. The results showed a statistically significant difference between EPIs and SA (in favour of EPIs), with EPI being beneficial in 73% of cases, compared with only 45% for SA. Patient positioning worsened with SA, especially in the longitudinal and vertical directions.

In a similar study, Zhao et al. (2016), used CT-on-rails (CTOR) as the reference standard in investigating surface alignment (SA), using the AlignRT system with CT-based target matching for pelvis treatment. All patients had multiple CTOR volumetric imaging as the gold standard reference for comparison with surface matching results. Ten patients were involved, and two CT simulation scans were acquired in the supine (alpha cradle) and prone (belly board) positions. On-treatment, initial laser patient positioning was followed by CTOR (90 degrees in-room) and SA (with the main detection region of interest (ROI) chosen not to include legs, to avoid day-to-day leg variations). Set-up errors were calculated for each, and differences were analysed for supine and prone patient positioning. Differences were found between results using CTOR and SA with the greatest difference shown, in prone rather than supine positioning. The overall conclusion of Zhao et al. (2016) was that SA was not a suitable image-guidance approach for pelvic treatments, for either supine or prone treatment positions, with potentially large discrepancies (up to 2.25 cm) found between the two methods.

As noted earlier in the chapter, ultrasound (US) has been used for many years for on-treatment image guidance for prostate patients. Consequently, Krengli et al. (2016) looked to compare SGRT (AlignRT) with US (Clarity) for prostate patients. The comparison here was with TAUS on 40 patients who had localised PCa, in a prospective study. All patients had appropriate bladder and bowel preparations. Both Clarity and AlignRT were installed in the treatment room. The overall conclusions from the study were that SGRT could match and predict set-up displacement detected and measured using TAUS, after adjustments for systematic errors. The two non-ionising methods were deemed complementary to one another, able to provide daily, low-cost, non-invasive, monitoring modalities for prostate patients intrafractionally, in addition to weekly X-ray based methods for prostate on-treatment image guidance.

Apicella et al. (2016) undertook a similar study to evaluate SGRT (AlignRT) for intrafraction motion monitoring during pelvis treatments. From consecutive surface alignment (SAl), continuous monitoring in 3D showed remarkable intra-fraction set-up variations during radiotherapy treatment delivery. Their study involved 29 pelvis patients (22 male (21 prostate), 7 female), treated with IMRT, over 22–33 sessions (fractions) for prostate patients. Delivery times were 10–20 mins for prostate patients, who were treated supine with leg immobilisation, a comfortably full bladder and empty rectum. The most likely cause of patient shift during treatment intrafractionally was considered to be contraction and relaxation of pelvic muscles, possibly related to patient discomfort after set-up. The authors felt that their study supported the use of 3D SAI for intrafraction monitoring, especially for longer-lasting treatments.

In more recent work, Mannerberg et al. (2021) compared SGRT with a three-point laser set-up for prostate SBRT patients, in order to examine the speed and efficiency of set-up using SGRT (surface imaging (SIm)) compared with traditional three-point laser positioning for SBRT UHF prostate treatments, and to determine any effect on overall set-up accuracy. Figure 4.3 shows a surface imaging example within the study using the C-RAD Catalyst system. Forty patients were included, all with localised PCa, with a 20/20 split between traditional and SGRT set-up. The C-RAD system was used, and set-up time was assessed through log files within the oncology management system (OMS) (ARIA). On-treatment geometric verification was achieved from orthogonal 2D kV images, using three gold FMs. Set-up deviations between planned and final set-up positions for treatment delivery were compared between the two patient-positioning methods (SIm vs three-point

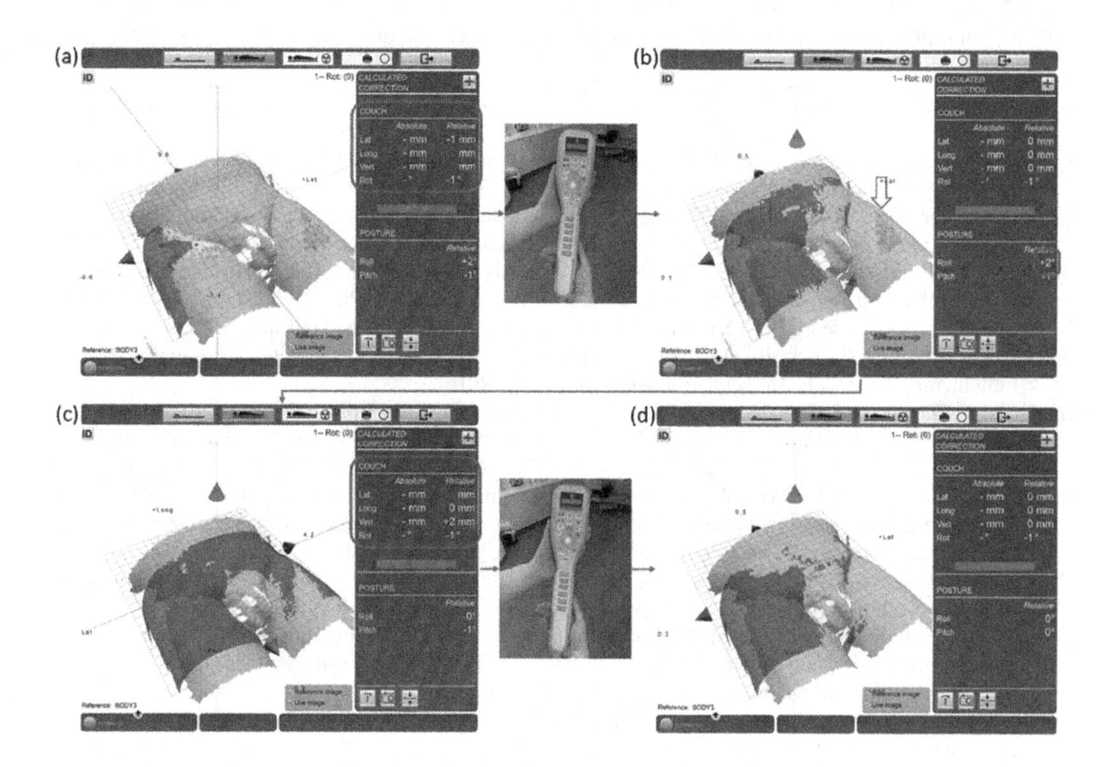

FIGURE 4.3 A surface imaging example within the Mannerberg et al. (2021) study comparing the surface guidance system with traditional three-point laser set-up for SBRT UHF prostate patients. The darker surfaces are the reference and lighter surfaces show the live surfaces, respectively. The couch was initially shifted to the isocentre position using saved couch parameters. The shift indicated by Catalyst™ (Lat –1 mm, Lng +10 mm, Vrt –7 mm) (a) was then applied using the Auto-GoTo function. The colour map and the positioning result indicated a roll (b), which was corrected for by asking the patient to adjust himself. Once the roll was corrected for, residual translations (Lat +4 mm, Lng 0 mm, Vrt +2 mm) (c) were applied using Auto-GoTo into the correct treatment position (d). Image taken from Figure 1 of Mannerberg et al. (2021). Faster and more accurate patient positioning with surface guided radiotherapy for ultra-hypofractionated prostate cancer patients. Technical Innovations & Patient Support in Radiation Oncology. 19:41–45. Published by Elsevier B.V., – an open access article under the CC BY Licence.

localisation). Treatment delivery was 42.1 Gy in seven fractions, using six megavoltage (MV) flattening filter free (FFF) volumetric modulated arc therapy (VMAT). Patients were planned with CTV-PTV margins of 8 mm, in line with the HYPO-RC-PC trial (see Chapters 8 and 9), with the CTV defined as only the entire prostate gland.

Results showed a statistically significant improvement (reduction) in initial patient positioning set-up time using SIm compared with the three-point localisation, with a median set-up time of 2:50 m:s (1:32–6:56 m:s) for SIm compared with 3:28 m:s (1:42–12:57 m:s) for three-point localisation. Set-up position was also found to be better for patients using SIm compared to those using the three-point localisation, with a median offset vector of 4.7 mm (0–10.4 mm) for SIm compared with 5.2 mm (0.41–17.3mm) for three-point localisation. In general, the authors considered SIm patient set-up time to be reduced by approximately 1 min per treatment, with an added improved initial patient set-up observed in those patients positioned using SIm compared with the traditional three-point localisation using lasers and skin marks, prior to IGRT for these SBRT UHF prostate treatments.

In terms of SGRT use in general, a recent international survey on behalf of ESTRO by Batista et al. (2022) gathered the responses from 278 international centres. They found that 172 had at least one SGRT system, 136 using it clinically. Many reported at least partial elimination of skin marks (58/135). Most combined SGRT with radiographic image-guidance patient positioning (115/136) and motion management (104/136), with 58% of the 136 respondents using it for pelvis treatments.

REFERENCES

Apicella, G., Loi, G., Torrente, S., et al. 2016. Three-dimensional surface imaging for detection of intra-fraction setup variations during radiotherapy of pelvic tumors. *La Radiologia Medica.* 121(10): 805–810.

Batista, V. Gober, M., Moura, F., et al. 2022. Surface guided radiation therapy: An international survey on current clinical practice. *Technical Innovations & Patient Support in Radiation Oncology.* 22: 1–8.

Berchtold, J., Winkler, C., Karner, J., et al. 2023. Noninvasive inter- and intrafractional motion control in ultrahypofractionated radiation therapy of prostate cancer using RayPilot HypoCath™—A substitute for gold fiducial-based IGRT? *Strahlentherapie und Onkologie.* Published online: 25 August 2023.

Bertholet, J., Knopf, A., Eiben, B., et al. 2019. Real-time intrafraction motion monitoring in external beam radiotherapy. *Physics in Medicine And Biology.* 64: 15TR01.

Biston, M-C., Zaragori, T., Delcoudert, L., et al. 2019. Comparison of electromagnetic transmitter and ultrasound imaging for intrafraction monitoring of prostate radiotherapy. *Radiotherapy and Oncology.* 136: 1–8.

Braide, K., Lindencrona, U., Welinder, K., et al. 2018. Clinical feasibility and positional stability of an implanted wired transmitter in a novel electromagnetic positioning system for prostate cancer radiotherapy. *Radiotherapy & Oncology.* 128(2): 336–342.

Camps, S., Fontanarosa, D., de With, P., et al. 2018. The use of ultrasound imaging in the external beam radiotherapy workflow of prostate cancer patients. *Biomed Research International.* 2018: 7569590.

Chaurasia, A., Sun, K., Premo, C., et al. 2018. Evaluating the potential benefit of reduced planning target volume margins for low and intermediate risk patients with prostate cancer using real-time electromagnetic tracking. *Advances in Oncology.* 2018(3): 630–638.

Colvill, E., Booth, J., Nill, S., et al. 2016. A dosimetric comparison of real-time adaptive and non-adaptive radiotherapy: A multi-institutional study encompassing robotic, gimbaled, multileaf collimator and couch tracking. *Radiotherapy and Oncology.* 119: 159–165.

D'Ambrosio, D. J., Bayouth, J., Chetty, I. J., et al. 2012. Continuous localization technologies for radiotherapy delivery: Report of the American Society for Radiation Oncology Emerging Technology Committee. *Practical Radiation Oncology.* 2: 145–150.

Dang, A., Kupelian, P., Cao, M., et al. 2018. Image-guided radiotherapy for prostate cancer. *Translational Andrology and Urology.* 7(3): 308–320.

De Los Santos, J., Popple, R., Agazaryan, N., et al. 2013. Image guided radiation therapy (IGRT) technologies for radiation therapy localization and delivery. *International Journal of Radiation Oncology, Biology, Physics.* 87(1): 33–45.

Foster, R. D., Pistenmaa, D. A., Solberg, T. D. 2012. A comparison of radiographic techniques and electromagnetic transponders for localization of the prostate. *Radiation Oncology.* 7: 101.

Ghadjar, P., Fiorino, C., af Rosenschold, P., et al. 2019. ESTRO ACROP consensus guideline on the use of image guided radiation therapy for localised prostate cancer. *Radiotherapy and Oncology*. 141: 5–13.

Gorovets, D., Burleson, S., Jacobs, L., et al. 2020. Prostate SBRT with intrafraction motion management using a novel linear accelerator-based MV-kV imaging method. *Practical Radiation Oncology*. 10: e388–e396.

Kirby, M., Calder, K-A. 2019. *On-treatment verification imaging: A study guide for IGRT*. Boca Raton, FL: CRC Press, Taylor and Francis Group.

Krengli, M., Loi, G., Pisani, C., et al. 2016. Three-dimensional surface and ultrasound imaging for daily IGRT of prostate cancer. *Radiation Oncology*. 11(1): 159–166.

Li, M., Hegemann, N-S., Manapov, F., et al. 2017. Prefraction displacement and intrafraction drift of the prostate due to perineal ultrasound probe pressure. *Strahlentherapie und Onkologie*. 193: 459–465.

Lovelock, D., Messineo, A., Cox, B., et al. 2015. Continuous monitoring and intrafraction target position correction during treatment improves target coverage for patients undergoing SBRT prostate therapy. *International Journal of Radiation Oncology Biology Physics*. 91(3): 588–594.

Mannerberg, A., Kugele, M., Hamid, S., et al. 2021. Faster and more accurate patient positioning with surface guided radiotherapy for ultra-hypofractionated prostate cancer patients. *Technical Innovations & Patient Support in Radiation Oncology*. 19: 41–45.

Olsen, J. Noel, C., Baker, K., et al. 2012. Practical method of adaptive radiotherapy for prostate cancer using real-time electromagnetic tracking. *International Journal of Radiation Oncology, Biology, Physics*. 82(5): 1903–1911.

Pallotta, S., Vanzi, E., Simontacchi, G., et al. 2015. Surface imaging, portal imaging, and skin marker set-up vs. CBCT for radiotherapy of the thorax and pelvis. *Strahlentherapie Onkologie*. 191(9): 726–733.

Panizza, D., Faccenda, V., Lucchini, R., et al. 2022. Intrafraction prostate motion management during dose-escalated linac-based stereotactic body radiation therapy. *Frontiers in Oncology*. 12: 883725.

RCR (Royal College of Radiologists). 2021. *On-target 2: Updated guidance for image-guided radiotherapy*. London: The Royal College of Radiologists.

Richter, A., Exner, F., Weick, S., et al. 2020. Evaluation of intrafraction prostate motion tracking using the Clarity Autoscan system for safety margin validation. *Medical Physics Journal*. 30(2020): 135–141.

Trainer, M., Adamson, S., Carruthers, L., et al. 2020a. Analysis of the intra-fractional motion of the prostate during SBRT using an EM Transmitter. *International Journal of Radiation Oncology, Biology, Physics*. 108(3):Suppl: E340.

Trainer, M., Nailon, B., Carruthers, L., et al. 2020b. Investigating the use of RayPilot for motion management during prostate SBRT: Initial experience. *Radiotherapy and Oncology*. 152:Suppl 1:S869–S870.

Vanhanen, A., Poulsen, P., Kapanen, M. 2020. Dosimetric effect of intrafraction motion and different localization strategies in prostate SBRT. *Physica Medica*. 75: 58–68.

Vanhanen, A., Syren, H., Kapanen, M. 2018. Localization accuracy of two electromagnetic tracking systems in prostate cancer radiotherapy: A comparison with fiducial marker based kilovoltage imaging. *Physica Medica*. 56: 10–18.

Wohlfahrt, P., Schellhammer, S. 2022. Chapter 4: In-room systems for patient positioning and motion control. In *Image-guided high-precision radiotherapy*. Ed. E. Troost. Cham: Springer Nature.

Willoughby, T., Kupelian, P., Pouliot, J., et al. 2006. Target localization and real-time tracking using the Calypso 4-D localization system in patients with localized prostate cancer. *International Journal of Radiation Oncology Biology Physics*. 65(2): 528–534.

Zhao, H., Wang, B., Sarkar, V., et al. 2016. Comparison of surface matching and target matching for image-guided pelvic radiation therapy for both supine and prone patient positions. *Journal of Applied Clinical Medical Physics*. 17(3): 14–24.

5 Change Response and Adaptive Radiotherapy

5.1 INTRODUCTION

Aspects of change response and adaptive radiotherapy for IGRT generally are introduced in the following sections of Kirby and Calder (2019), to which the reader is directed first as a refresher or first-time read before engaging in the material in the current chapter.

- Chapter 7 – especially sections 7.5, 7.6 and 7.7
- Chapter 8 – sections 8.3–8.6
- Chapter 10 – for alternative technologies including MR (section 10.7) and Adaptive Radiotherapy (section 10.10)

This is an area that is undergoing much research, not least for prostate cancer (PCa) patients, examining new magnetic resonance (MR)-guided and computed tomography (CT)-guided methods of adaptive radiotherapy (ART), and also investigating ways of understanding, monitoring, minimising the effects of or tracking/compensating for intrafractional motion, especially for the newest ultra-hypofractionated (UHF) treatment regimes which are now showing their clinical effectiveness in localised PCa treatments.

Key books and papers which can give the reader a good grounding in ART and change response in general and for PCa in particular are RCR (2021, pp. 63–74); Kirby and Calder (2019, chapter 10); Wohlfahrt and Schellhammer (2022); Hackett et al. (2022); Zou et al. (2018); Gregoire et al. (2020); Pathmanathan et al. (2018); McPartlin et al. (2016); Kashani and Olsen (2018); Sonke et al. (2019); and Bertholet et al. (2019).

5.2 MR-BASED SYSTEMS AND METHODS

MR imaging is already an important tool within radiotherapy (and especially for prostate cancer (PCa) patients) for its soft tissue visualisation and unique tissue information. For MR imaging for radiotherapy purposes, one requires equipment as used for treatment delivery – e.g., flat, indexed couch tops, a large-bore machine for accommodating immobilisation and scanning in the treatment position, positioning lasers, surface coils (for improved signal-to-noise (SNR)) but applied without affecting body contours and patient anatomy important for correct dose planning and delivery, appropriate distortion correction, artefact reduction and high SNRs, etc. (Zou et al. 2018; McPartlin et al. 2016).

In aiding pre-treatment, MR imaging (fused with CT or in newer MR-only workflow environments (Bird et al. 2019; Speight et al. 2021)) offers the advantages of better-contrast soft tissue imaging for more accurate target and organ at risk (OAR) delineation, thereby reducing interobserver variations and offering the ability to reduce margins and treat smaller volumes with greater confidence (although this then depends upon good on-treatment image guidance). Multiparametric MR imaging (mpMRI) offers higher sensitivity and specificity in PCa patients, especially in terms of identifying potential dominant intraprostatic lesions (DILs) and the option to focal boost (Zou et al. 2018; McPartlin et al. 2016; Ma et al. 2021, 2023; Ladbury et al. 2023).

For on-treatment image guidance and adaptive radiotherapy (ART), there are various designs and methods of MR LINACs (Kirby and Calder 2019, chapter 10; Sritharan and Tree 2022), for

DOI: 10.1201/9781003050988-5

FIGURE 5.1 A photograph of the new Elekta Unity MR LINAC, one of the first to be commercially available. Image courtesy of Elekta.

better soft tissue visualisation and 4D cine-MR imaging intrafractionally, with no concomitant dose burden. MR-only workflow pre-treatments are methods for dose calculations on MR datasets alone, synthetic CT generation (Johnstone et al. 2018), which will then naturally allow propagation into dose calculations and optimisation on-treatment for ART and eventually dose-guided radiotherapy. MR LINACs are still being developed but are now a commercial reality (see Figure 5.1), offering opportunities for correcting translations, rotations and deformations of the prostate and seminal vesicles (SVs) in real time throughout treatment, with both inter- and intrafraction corrections feasible through this technology (Ghadjar et al. 2019), although, for PCa patients, its full benefit is still being researched (see Chapters 8 and 9). On-treatment MR LINAC challenges are still present (asymmetric penumbra, electron return effects, etc.), but solutions are being presented in development (Zou et al. 2018; McPartlin et al. 2016; Acharya et al. 2016; Sritharan and Tree 2022; Kirby and Calder 2019, chapter 10).

Through advanced image-guided technologies like MR LINACs, opportunities exist for a safe decrease in margin sizes with accompanying reductions in OAR volume irradiation and therefore potentially lower toxicities, although this is also dependent on the overall dose and fractionation regime (Ma et al. 2021, 2023). It also brings the capability of visualising and tracking targets in real time and developing online ART, all with zero concomitant dose burden. Clinical trials (see Chapters 8 and 9), like PRISM (Pathmanathan et al. 2018; Sritharan and Tree 2022), MIRAGE (Kishan et al. 2023), HERMES (Westley et al. 2023) and SMART (Bruynzeel et al. 2019; Leeman et al. 2022), to name but a few, are demonstrating safety, implementation, effectiveness, initial clinical outcomes and the ability to safely use tighter margins using this image guidance for PCa patients. The technology brings with it possibilities for dose escalation, where necessary, without increased side effects (Sritharan and Tree 2022; Tocco et al. 2020).

But initial designs and implementation also present challenges, like higher resource issues (staffing and costs), technique capabilities, etc., which will maintain a distance between its use for full ART and perhaps its full potential and widespread replacement of current radiotherapy equipment and methods.

Other image-guided modalities (some as detailed in this and other chapters in this book) bring their own capabilities, advantages and challenges, but can incorporate real-time intrafractional tracking and gating, such as robotic arm systems, triggered imaging, X-ray-based enhanced faster imaging on O-ring gantry systems, Artificial intelligence (AI)-aided ART methods on X-ray images, etc., all of which can also enable the use of tighter margins alongside ultra-hypofractionated treatment regimes. Trials such as PACE-B (Van As et al. 2023) and 2SMART (Ong et al. 2023), for example, demonstrate possibilities and achievable outcomes with X-ray-based image guidance for PCa patients (see Chapters 8 and 9). Although comparisons are already being presented comparing MR guided (MRg) and CT guided (CTg) for PCa patients (Leeman et al. 2022), further evidence is still required, especially at level 1, to fully compare and contrast the two modalities and present the advancing developments for PCa patients through such image guidance.

Siritharan and Tree (2022) also note the challenges that are still facing this new technology and its use, such as geometric image distortion, artefacts in MR images, balance of speed of acquisition and image quality, the electron return effect (Kirby and Calder 2019, chapter 10), lack of non-coplanar and electron beam therapy, bore size and treatment beam dimensions (especially cranio-caudally – max 22cm, making 80% of plans feasible with 1 cm margin, so all prostate patients could be treated), longer treatment times (especially for ART due to recontouring and reoptimisation and quality assurance (QA)), noise and difficulties for claustrophobic patients (e.g., up to 1 hour compared with 10 mins for standard intensity modulated radiotherapy/volumetric modulated arc therapy (IMRT/VMAT) and 20–30 mins for X-ray cone-beam computed tomography (CBCT)-based Ethos ART (see Section 5.3). Research shows that 5% of patients find this to be lengthy (most patients find this OK, but the most frequent complaints are noise, paraesthesia and cold). There is a need for an multidisciplinary team (MDT) for daily online adaptive treatments (initially a large team and for the first implementation only seven patients per day; see Section 5.6). One must look for the MR LINACs to move from research to routine clinical implementation, for which resources (including training; see Chapter 11) are an issue, not least in MR safety, MR anatomy and image interpretation, cost and resource burden, difference in maintenance and quality control (QC) compared with standard X-ray-based image-guided treatment units, etc. (Sritharan and Tree 2022; Kirby and Calder 2019, chapter 10).

The benefits of superior image quality are not to be underestimated. This will help reduce interobserver variability, especially at the difficult-to-delineate (but crucial) soft tissue boundaries between the prostate and the rectum/bladder (Murray and Tree 2019; Ladbury et al. 2023). With better, more accurate delineation comes the possibility of using smaller margins, with verification and adaptation on-treatment through the MR LINAC and therefore reduced high-dose volumes and lower toxicities (Sritharan and Tree 2022, Acharya et al. 2016).

Current set-up corrections (CBCT and CT-based) allow for visualisation of internal anatomy but offer correction before treatment delivery starts; technology is only just beginning to offer online ART solutions based on improved quality CBCT images (see Section 5.3). On most C-arm LINACs, CBCT IGRT involves only positional corrections via couch movements – it does not offer a response to anatomical geometrical changes, which can happen for the prostate, especially with respect to seminal vesicles (SVs) and pelvic lymph nodes (PLNs) when involved in treatment, and which move independently of the prostate, although ART through X-ray-based systems is changing the landscape here (see Section 5.3). Previous research has shown that up to one-third of fractions delivered to PCa patients would warrant modification when comparing the original plan with CBCT anatomy due to differences in delivered dose distribution compared with the original planned one. Online ART enables modification of the plan to suit the anatomy of the day, accounting for interfractional changes in anatomy by daily adaptation of contours (target and OAR) and replanning, compared with rigid body style corrections and/or imaging with FMs (which will not necessarily account for all shape and positional changes of the target volume). These markers themselves require an invasive procedure, something that is not required for MR-only workflows pre-treatment when on-treatment image guidance is MR based (McPartlin et al. 2016; Zou et al. 2018; Sritharan and Tree 2022).

MR LINACs also bring forward the possibilities of real-time cross-sectional imaging during the treatment delivery (Kashani and Olsen 2018). For most CBCT methods, including current X-ray-guided ART, imaging is acquired before treatment delivery, but other methods for motion monitoring and tracking during treatment delivery in real time exist (see Section 5.4).

Prostate motion intrafractionally is well known. Non-periodically, the prostate itself can move independently and/or because of bladder filling and rectal motion (e.g., through movement of rectal gas). Movement has the potential to impact dosimetric coverage and (especially with tight margins and highly conformal dose distributions) doses to OARs such as the bladder and rectum themselves. Some studies show movements greater than 3 mm in 10–15% of PCa patients; more recent data using on-treatment MR imaging show that the movement can be less, but some still have larger displacement, for whom the dose delivered will be noticeably different from that planned. Monitoring and tracking may help with these patients with MR LINACs in the future (Ghadjar et al. 2019; Zou et al. 2018; Sritharan and Tree 2022).

These benefits become more pertinent for hypofractionated treatments (especially ultra-hypofractionated ones; see Chapter 8), treatment regimes where recent results demonstrate that they can likely become the standard of care for certain PCa patients (Loblaw 2020). With fewer fractions and higher dose per fraction, smaller margins with an accompanying precision of delivery become of heightened importance.

Future development will bring MR-only workflows, withno CT involvement; associated reduced costs, fewer patient visits and reduced concomitant dose to the patient (Bird et al. 2019). There could likely be fewer differences in contouring and reduced uncertainties from less need for CT-MR image fusion, which are usually caused by anatomical differences (e.g., bladder status) between MR and CT scans (Bird et al. 2019; Sritharan and Tree 2022). A key step for the MR-only workflow is production of synthetic CT for electron density information for dose calculations. Challenges with distortion and geometric accuracy are being overcome for pelvic patients, and the workflow has been shown to be feasible and clinically acceptable (Bird 2019; Johnstone et al. 2018).

Real-time imaging intrafractionally will become a possibility, likely alongside other motion management strategies (see Section 5.4), enabling pausing treatment when internal organ movement moves the target outside predefined tolerances or which risks OAR movement into higher dose volumes (through peristalsis, air in the rectum, breathing motion, etc.). Ultimately this could lead towards true intrafractional adaptation – where the treatment plan is adapted in real time with respect to detected motion. This would likely be most useful for ultra-hypofractionated (UHF) regimes, with consequent expected further reductions in toxicity (Pathmanathan et al. 2018).

MR-guided ART (MRgART) also offers the possibilities of re-irradiation of tumours with reduced margins and also the possibilities of monitoring better change response throughout the treatment course, and responding accordingly (dose response studies and dose-guided RT (see Kirby and Calder 2019, chapter 10). The possibilities for functional imaging are also there (Kirby and Calder 2019, chapter 10) to monitor response, differentiate post-therapy changes from active tumours and detect recurrent tumours (Sritharan and Tree 2022). Functional methods, like Diffusion Weighted Imaging (DWI) or monitoring parameters like the Apparent Diffusion Coefficient (ADC) as a biomarker have been shown to reveal interesting characteristics of tumours and their response to treatment. For example, for PCa patients, changes in ADC values during therapy have been linked to clinical outcomes. Although not without its limitations, this type of biomarker monitoring becomes more possible with the greater use of MR within RT, and on the MR LINAC. Acquired using the latter as part of the daily online ART workflow, daily response assessment may be possible, which could be used to modify the treatment plan and escalate dose to regions where the diffusion coefficient is lower, which may be where there are more radioresistant tumour cells. The full clinical

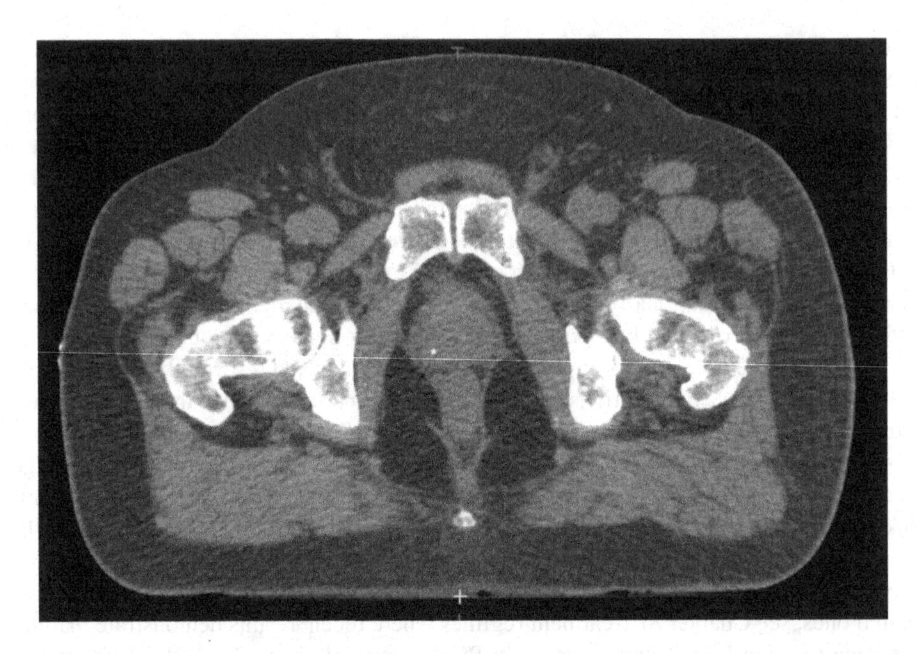

FIGURE 5.2 Image quality comparisons for (a) kV CT, (b) kV CBCT and (c) MR images for the prostate, showing the softtissue imaging improvements which are possible with MR imaging. The MR Image acquired on the Elekta Unity MR LINAC at the Royal Marsden NHS Foundation Trust. Images (d), (e) and (f) are enlarged sections of images (a), (b) and (c) respectively. Images courtesy of Prof. Helen McNair, The Royal Marsden NHS Foundation Trust, and the Institute of Cancer Research, London, UK.

value needs much further exploration and research, but signals some exciting future possibilities (Sritharan and Tree 2022; Ladbury et al. 2023; Zou et al. 2018; McPartlin et al. 2016; Pathmanathan et al. 2018; Kirby and Calder 2019, chapter 10).

For PCa patients in particular and their radiotherapy treatment, the imaging improvements for soft tissue are striking when compared with CT and CBCT (Figure 5.2). The prostate itself is seen more clearly on MR compared with other imaging modalities; mpMRI can give extra information on intraprostatic disease, increasing specificity and sensitivity for diagnosis with studies showing that it is more sensitive than transrectal ultrasound (TRUS)-guided biopsy for initial diagnosis. European guidance suggests mpMRI used for all patients as the primary modality for staging local disease and for target volume delineation, reducing interobserver variability in contouring the prostate and the important soft-tissue boundaries with the rectum and bladder (Murray and Tree 2019; Sritharan and Tree 2022; Ladbury et al. 2023).

A practical challenge is the one of MR in radiotherapy departments and the need for backup plans (and fiducial markers (FMs)) if patients cannot be treated on an MR LINAC. From the earliest clinical implementation, opportunities have been obvious; in being able to contour dose plans more precisely to daily changes in anatomy around the prostate and being able to target the dominant intraprostatic lesion (DIL) with much greater visibility, with an extra focal boost dose (Murray and Tree 2019).

The move to UHF clinical trials has been continuing for a few years now, and also with the delivery of these regimes on MR LINACs, with impressive early and beyond-five-year results on toxicity and control (see Chapters 8 and 9). Further progress is continuing, investigating the possibilities now of fewer than five-fraction treatments, on both MR LINACs and conventional LINACs through the HERMES (Westley et al. 2023) and 2SMART (Ong et al. 2023) trials.

One of the most important aspects of online ART using the MR LINAC is the workflow adopted. From the earliest clinical systems, Acharya et al. (2016) published their first clinical experiences with the Viewray MRidian system (Kirby and Calder 2019, chapter 10). The basic principle is to

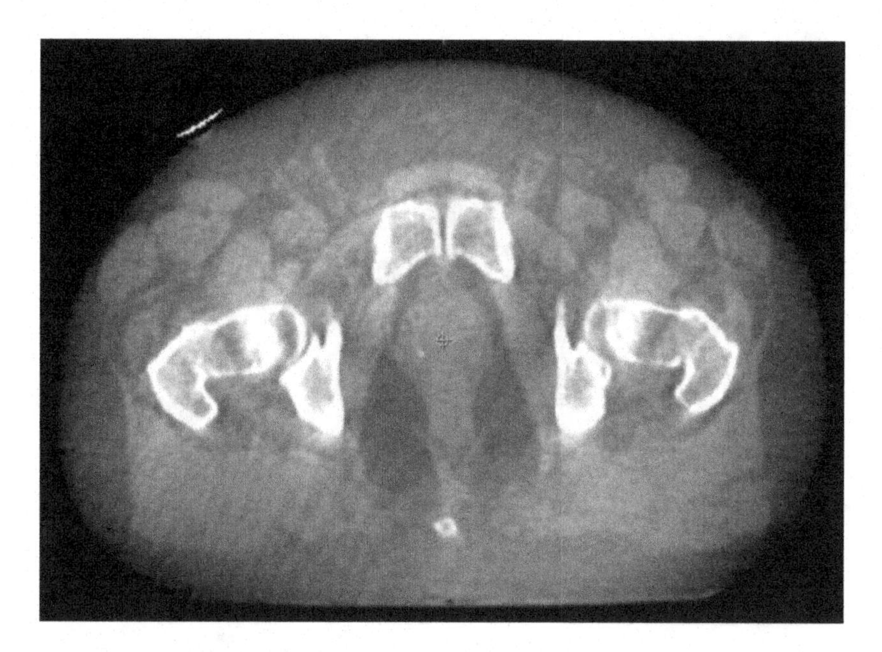

FIGURE 5.2B Continued

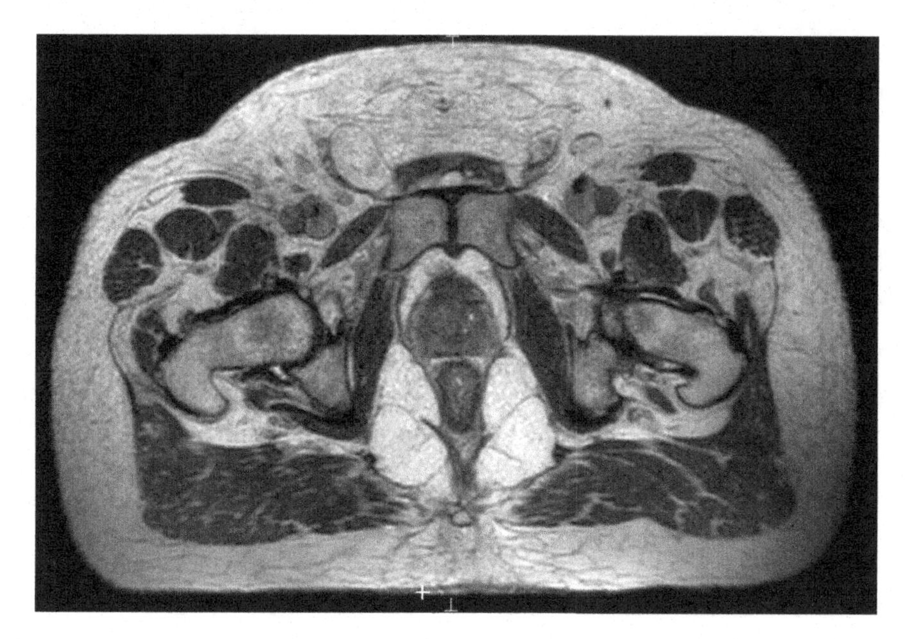

FIGURE 5.2C Continued

use the anatomy of the day from the MR image to inform the adaptation process. One of the first approaches devised and published by Acharya et al.'s (2016) first clinical applications introduced the use of deformable modelling and registration to transfer planning CT contours and electron density information onto the daily MR scan. This is performed online with the patient on the couch, after the initial MR scan with the MR LINAC and is used to recalculate the dose distribution onto the daily MR image for clinical evaluation during the adaptation process.

If required, the contours are adjusted and the dose is then calculated (from the original plan) on the MR image of the day. If deemed acceptable, then the original plan is used for treatment. If not, then the plan is reoptimised with the new contours to produce a new plan for the day. The plan

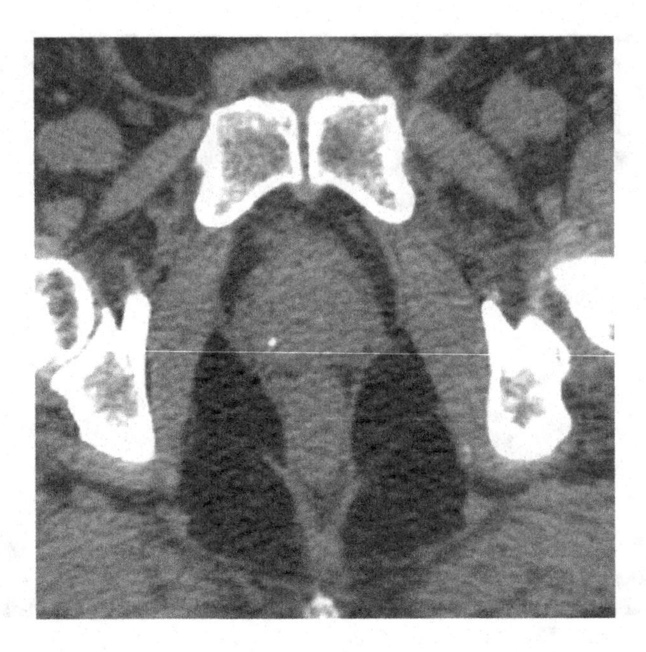

FIGURE 5.2D Continued

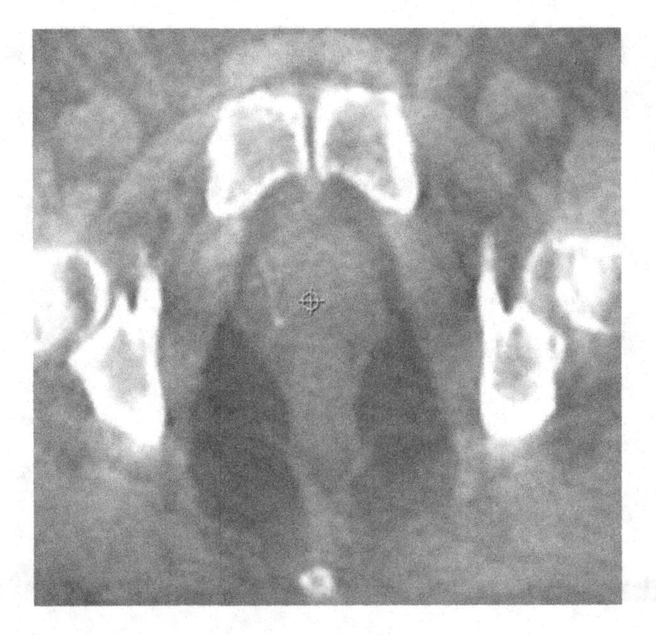

FIGURE 5.2E Continued

is then compared with the original plan and the better plan is selected. If this is the original plan, treatment commences. If not, then the new plan has patient-specific QA applied before the treatment is then delivered, if all is satisfactory (Acharya et al. 2016). When first implemented, the median time for recontouring, reoptimisation and QA for the first five patients (all with abdominopelvic tumours); then with a larger cohort (20 patients including the first five) was 26 mins. For the larger patient cohort (20 patients, 170 fractions), the adapted plan was never worse than the original plan, and was used in approx. 55% of fractions. Most patients (63%) had a total online ART time of less than 30 mins (Acharya et al. 2016).

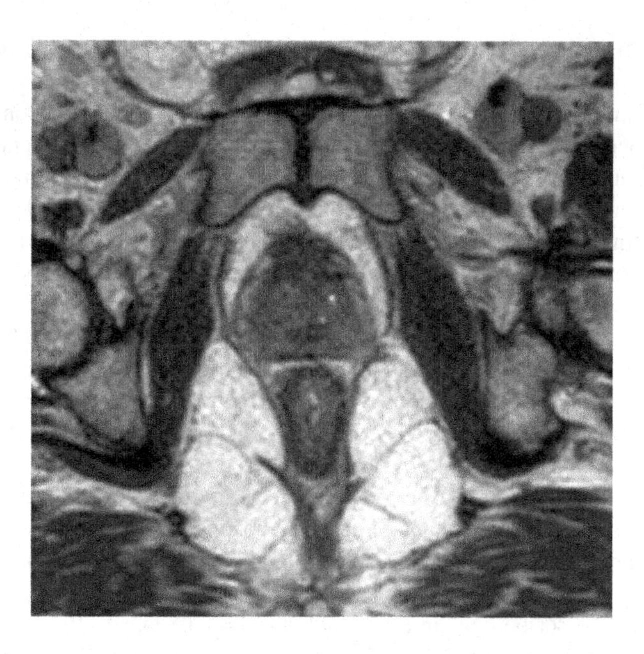

FIGURE 5.2F Continued

The workflow on the Elekta Unity MR LINAC system is similar (Winkel et al. 2019), but was initially divided initially into two processes – one named "Adapt-to-Position" (ATP) and the other named "Adapt-to-Shape" (ATS). Each process begins with an acquired MR image. Pre-treatment CT, contours and plan, together with the online MR scan, are used as input to the adaptation process. The two different workflows both use the Monaco treatment planning system (TPS).

The ATP workflow works primarily on patient-position-based adaptation. The pre-treatment CT is matched with the MR image using rigid registration, and the isocentre position is updated on the reference image and dataset. The pre-treatment plan is recalculated or reoptimised to reproduce or improve further target coverage by a library of different adaptation methods (which increase in complexity from using original segments to adapted ones to reoptimised weights and shapes). This reoptimisation is on the pre-treatment CT and contours, so original contours are used (Winkel et al. 2019).

The ATS workflow works with a new anatomy-driven plan adaptation, where optimisation is with the daily MR and adapted contours (Winkel et al. 2019). Pre-treatment CT and online MR are first registered, as with the ATP workflow. Contours are then propagated automatically through deformable models onto the online MR. Editing of contours is undertaken, if deemed necessary, and electron densities are applied derived from the average values within structures on the pre-treatment CT. This recalculation/reoptimisation is therefore conducted on the online MR scan and new contours. Like the ATP workflow, multiple options are available for the recalculation/reoptimisation process (Winkel et al. 2019).

Tocco et al. (2020) examined their MR-guided ART (MRgART) approach for PCa patients, comparing MR-guided ART for both the Viewray MRidian and Elekta Unity workflows, as a joint paper from institutions having these two main commercial systems, with workflows, experiences, resource implications, challenges and opportunities being compared and contrasted with MRgART for PCa patients. Training and workforce resources are highlighted as challenges, together with the logistics of equipment availability for more routine use. For PCa patients, who may be at an age where there are other medical contraindications, some of the difficulties with MR scans are mentioned, such as MR screening necessity (as for a normal diagnostic MR scanner), issues with noise, claustrophobia, etc. (Tocco et al. 2020).

Lawes et al. (2022), in recently published work, discussed how the workflows on the Elekta Unity system could be adapted further, and additional MR images acquired immediately prior to treatment delivery, to examine the effects of any intrafractional motion during the long time component of the adaptation process. The purpose was to develop further, simpler adjustments made to account for this motion without necessarily re-performing a full adaptation and thereby increasing session time considerably. This process itself can be refined further by verifying if the visible prostate is within a corresponding planning target volume (PTV) and therefore allowing an appropriate decision as to whether to extend the session time further, for what might be little dosimetric gain (Lawes et al. 2022)

Poon et al. (2022a) described their experience with the Unity MR LINAC for five-fraction stereotactic body radiotherapy (SBRT) treatments for 107 PCa patients and retrospectively analysed over 500 fractions of treatment. In total, for approx. 16% of fractions, the ATS workflow was used, and for the remaining 84% of fractions, the ATP workflow was used. For some treatments, there was a distinct change in the ATS to ATP workflow as patients progressed from the first to the last fraction; as different (template-based) criteria were applied each time to assess the need for an ATS-type workflow. These assessments included anatomical and dosimetric criteria, as well as QA and other miscellaneous aspects. In most cases, it was solely the anatomical criteria that guided the decision to use ATS (Poon et al. 2022a).

Dassen et al. (2023) describe how an additional workflow (Adapt-to-Rotation (ATR)) has been introduced to balance speed and accuracy within the MRgART process, in order to correct for translations and also rotations. This study, involving 26 patients with intermediate-/high-risk PCa treated within the hypo-FLAME 2.0 trial (Draulens et al. 2020; see Chapters 8 and 9), looked to quantify differences in accuracy of online adaptive workflows by evaluating the margins needed to accommodate intrafractional motion estimated for the prostate and the SVs in each of the workflows on the Elekta Unity MR LINAC. Following the adaptation process, before beam-on, another MR scan was acquired (called a position-verification MRI) to assess intrafractional motion; when compared with the initial MR scan used for adaptation, if observed shifts equalled or exceeded 2 mm of the clinical target volume (CTV), then a shift was applied before treatment commenced. Halfway through the beam-on time, a further MR scan was acquired to aid with assessing intrafractional motion with respect to each of the ATP, ATR and ATS workflows. Timelines are illustrated within the paper of Dassen et al. (2023), with mean treatment times given of approximately 25 mins, 29 mins and 55 mins for the ATP, ATR and ATS workflows, respectively, for these SBRT five-fraction PCa treatments. Results showed that the margins used were appropriate to maintain coverage of the CTV; for some cases, a reduction in margin could have been used (by 0.5–1.5 mm) with the ATS workflow, but one must weigh the balance of advantages and resource issues for the longer ATS process (Dassen et al. 2023).

While clinical work is ongoing through the limited number of MR LINACs available, *in-silico* studies are still being performed to further inform our ART treatments for PCa patients. Here, Christiansen et al. (2022) used an *in-silico* planning study to investigate the potential PTV and toxicity reduction which might be possible through the use of ART on MR LINACs. Two cohorts of patients were compared: one treated on the Elekta Unity MR LINAC, using moderate (20 fraction) and UHF (seven fraction) regimes and ATS protocols for localised PCa; and one on standard LINACs with a conventional VMAT 2 Gy per fraction regime to the prostate (78 Gy) and PLNs (56 Gy) for high-risk disease. MR scans were acquired at three points for the conventionally treated patients to simulate MRgART for the high-risk patients. Both cohorts were compared using dose volume histogram (DVH) analysis and normal tissue complication probability (NTCP) models. Although validation should ultimately come from a clinical trial, the results showed that, if daily MRgART was applied for the high-risk patients, PTV margins could be reduced, potentially reducing acute GI toxicity as well as late bladder toxicity, through decreased doses to the peritoneal cavity, rectum and bladder (Christiansen et al. 2022).

Turkkan et al. (2022) have described their experience with using the Elekta Unity MR LINAC for PCa patients treated with conventional, moderate and UHF regimes. Fourteen patients were involved in total, receiving a total of 375 fractions; five patients (36%) were treated conventionally, six (42%) were treated with a moderate regime and three (22%) with a UHF regime. Most fractions (372) were treated using an ATS protocol.

Total fraction time was found to decrease in the second three months of the study compared to the first, from a mean of 48 mins (40–81) down to 43 mins (24–72). This was attributed partly to experience and partly to software and hardware changes made during the trial. Machine-related interruptions in 11/375 fractions (3%) decreased to just under 1% after the upgrade. No grade 3 toxicity, or worse, was reported, although some Grade 1/2 genitourinary (GU) and gastrointestinal (GI) toxicities were reported.

Bruynzeel et al. (2019) and Tetar et al. (2021) have reported their findings on a prospective single-arm phase 2 study for MRgART on the Viewray MRidian system, examining early toxicities after extreme hypofractionated SBRT treatments for localised PCa. The second paper reports results focussing on patient-reported outcomes (PROs). One hundred patients were enrolled; all but four with intermediate-/high-risk disease. Doses of 36.25 Gy in five fractions with daily online plan adaptation were prescribed with plans enabling simultaneous sparing of the urethra to a dose of 6.5 Gy per fraction. Early toxicity profile maximums for ≥ Grade 2 GU and GI toxicities were approximately 24% and 5%, respectively; no early Grade 3 or worse toxicities were reported.

Tetar et al. (2021) then reported further on the PROs from the study, examining health-related quality of life (QOL) characteristics at time points from the last fraction up to 12 months post treatment. Response was good (95%) at all study time points. For both the patient responses and also physician-assessed adverse events, the greatest effects appeared to be on urinary and bowel symptoms recorded in the first six weeks. Symptoms decreased after this time point, returning to baseline levels at 12 months. No Grade 3 or greater toxicities were reported and, in terms of limitations, no patients reported any due to urinary side-effects, while 2% reported some impact on daily activities due to bowel problems at one year. The results show that the treatment is well tolerated with low toxicity and only transient early urinary and bowel symptoms which resolved within 12 months.

More recently reported studies (for the MIRAGE and HERMES trials mentioned earlier) and their initial results can be found in Chapters 8 and 9. Further excellent studies and reviews to consult are the papers by Ladbury et al. (2023) Leeman et al. (2022, see chapter 9), who systematically review and perform a meta-analysis on published data comparing MRgART against FM or CT-guided non-adaptive treatments for PCa patients in terms of acute toxicities.

Teunissen et al. (2023) have published the first results from the MOMENTUM study; here, the patients treated on high-field (1.5 T) MR LINACs for patients treated with UHF (5 x 7.25 Gy regimes) for low-, intermediate- and high-risk localised disease. A total of 425 patients were identified in the international registry (approx. 11% low, 82% intermediate and 7% high risk), with follow-up up to 12 months. Peak GU and GI toxicity was reported at three months with approx. 19% (GU) and 2% (GI) of patients demonstrating Grade 2 or above toxicity. For the non-ADT patients, both physician- and patient-reported erectile dysfunction seemed to worsen significantly between baseline and 12 months. Compared with the baseline, no relevant deterioration in patient-reported bowel and sexually active domains were observed at three, six and 12 months post-treatment; but significant reductions in urinary domain scores were noted at the same time points, and for sexual function domain scores at six and 12 months. So far, the results indicate a safe and effective treatment using the MR LINAC; with good information for helping and supporting patients through expected side effects following treatment.

5.3 CT-BASED SYSTEMS AND METHODS

Although there are distinct advantages to ART performed using MR-guidance on the treatment machine, outlined above, there are also distinct challenges involved. One must not forget, therefore, that X-ray based ART solutions are under constant development, not least for PCa patients. With continued improvements in X-ray volumetric imaging (particularly in terms of speed of acquisition on O-ring rotational therapy units and with image quality improvement (such as iterative CBCT) (Kirby and Calder 2019, chapter 10)), ART using X-ray based solutions is producing very good results, despite the clear differences with MR of always delivering some amount of concomitant dose and having underlying physics limitations in terms of X-ray subject contrast (Kirby and Calder 2019, chapters 5, 7 and 10).

Antico et al. (2019) quite recently published results from an interesting phantom and *in-silico* proof-of-principle study to examine the potential of a library plan approach to X-ray based CT ART for PCa. IMRT-based treatment delivery was used but with beam selection based upon current target volume position, evaluated through a dosimetric comparison with the intended treatment plan over a multiple fraction simulated treatment course, including inter- and intrafractional set-up deviations. The method performed better than a standard plan with set-up correction, improving the coverage of the treatment volume and reducing doses to the nearby OARs. They proposed that it might be particularly beneficial for hypofractionated IMRT treatments, where a higher precision is required for each fraction, to ensure that the high dose volume conforms to the target volume, with reduced margins which are needed, especially for future UHF techniques.

Very recently, Lavrova et al. (2023) published a review of CT-based ART techniques for RT in general, comparing them with MRgART. The most recent techniques are online adaptive techniques, with the adaptation and reoptimisation of the treatment plan taking place post on-treatment image acquisition, while the patient is on the couch. Current studies show that such processes are resource and time intensive, but they are improving through the use of artificial intelligence (AI) and other efficient software and hardware. In addition, they are improving mainly through experience and training in terms of workflow and staff resource. X-ray-based kV-CBCT is the most widely used imaging modality for developing this work, while trying to always optimise and improve aspects, such as poorer quality, speed of acquisition and concomitant dose burden.

Offline ART looks at a process of treatment replanning, based upon anatomical changes from the original pre-treatment scan and plan, but between fractions. Online adaptive methods aim to address the unpredictable, non-periodic motion and internal anatomy changes encountered with PCa treatments from one fraction to the next, but online, immediately before each treatment fraction. With both approaches, the desire is for decreasing toxicity and improving target coverage, adapting to anatomical changes visible through kV-CBCT on-treatment imaging (Lavrova et al. 2023).

The Varian Ethos system (based upon the latest Halcyon O-ring rotation therapy unit (see Figure 5.3) with very fast kV iCBCT imaging (Kirby and Calder 2019, chapter 10) (see Figure 5.4) and now using a new and larger-format flat panel imager (Hypersight) (Liu et al. 2023), is perhaps the most advanced available at this time, based on the rapid and improving image quality of X-ray based kV-CBCT images. Improved CBCT acquisition speed and image quality (through iterative reconstruction) (Kirby and Calder 2019, chapter 10; Liu et al. 2023) pave the way for improved soft-tissue delineation and therefore the increased success of both manual and AI-based autosegmentation and autocontouring methods, involving more successful deformable image registration algorithms and dose calculation and reconstruction methods, which can be within 2–3% of planning models based on the usual fan-beam kV CT imaging (Lavrova et al. 2023; Liu et al. 2023). Coupled with contour propagation from the scheduled treatment plan and with the possibility of multi-modality image access availability on-treatment, the on-line adaptive process is being clinically implemented with very good results. Although prone to motion artefacts, the increased speed of acquisition and, for sites like the prostate, where movement is slower and more limited, means that image quality is considerably better than earlier CBCT techniques (Liu et al. 2023).

Initial results, through either Ethos simulations using its emulator on retrospective data, *in-silico* studies or through early clinical implementation, already show dosimetric differences compared with continued use of the scheduled treatment plans with no adaptation, usually with improved target coverage, restored to near optimal levels in over one-third of fractions in some studies (Lavrova et al. 2023; Liu et al. 2023; Moazzezi 2021; Oldenburger 2023).

Resources (staff and time) are of concern for these X-ray-based ART solutions (as they are for MRgART) (see Section 5.6) – with early clinical implementation requiring a different and expanded staff complement at the treatment control desk throughout adaptive treatments for online tasks and decision making, while the patient is on the couch (Zwart et al. 2022; McComas et al. 2023; Byrne et al. 2022). But initial timings (Lavrova et al. 2023) report that contour propagation and manual contour adaptation can take about 8 mins, with a total time for patient set-up to post-treatment CBCT verification of about 28 mins. Other timing results are being reported by Liu et al. (2023), Byrne et al. (2022), Morgan et al. (2023) and Zwart et al. (2022), as seen later in this section.

In comparison with online MRgART, image quality will always be superior with MR imaging on the MR LINAC technologies. Both imaging modalities suffer from imaging artefacts – e.g., motion and distortion for MR imaging and hardening, motion, scatter, aliasing, etc. for X-ray CBCT. Quantitative and functional imaging is possible with MR, although dual energy techniques bring in new possibilities for CT technologies. Field of view can be more limited on CBCT compared with MR, the latter also offering the possibilities of fast intrafractional imaging, with zero dose burden (which is not possible with gantry-mounted on-board CBCT). Image acquisition times with MR are long with the latest CBCT technology; Liu et al. (2023) quoted the new Varian Hypersight with a 6-second rotational acquisition time. Dose burden for modern X-ray techniques is of issue; Lavrova et al. (2023) quoted studies reporting up to 10 cGy per scan, but with typical values closer to 0.2–2 cGy.

Overall equipment costs are high for both technologies compared with standard IGRT-enabled C-arm LINACs, but the MR LINAC costs currently are higher than, for example, O-ring gantry-based CBCT online ART equipment. Current reports suggest a longer overall treatment duration for online MRgART compared with CBCT-based equipment and processes for PCa treatments (Zwart et al. 2022; Liu et al. 2023; Tetar et al. 2021).

Training may be considered a greater and additional requirement for MR LINACs compared with CBCT ART systems (Liu et al. 2023; see also Section 5.6 and Chapter 11). Human resources are of concern, in terms of new working patterns and staff requirements, considering the necessity for extra staff (of different disciplines) needed on-treatment compared with current IGRT on standard C-arm LINACs with just a complement of therapeutic radiographers (Lavrova et al. 2023).

Contouring, reading and evaluating MR is different than CT and requires more specialist training. It can be argued that radiotherapy staff are familiar working with CT and CBCT cross-sectional and volumetric imaging, MR is likely to require further training and experience. This is likely more an issue for manual contouring, but increases for both technologies in terms of verification and decision-making processes to ensure the accuracy and integrity of autosegmentation and autocontouring before reoptimisation and creation of adapted plans (Lavrova et al. 2023; Liu et al. 2023; Tocco et al. 2020).

Mention has been made earlier of Ethos emulators, hardware and software which can use clinical data to emulate the Ethos AI-driven software process (Archambault et al. 2020) but offline in a non-clinical environment. Morgan et al. (2023) retrospectively studied (*in-silico*) data from previous patients (pre- and post-treatment CBCT for post-operative patients), who were treated with wide margins to account for daily set-up errors and changes in internal anatomy and therefore could benefit from ART and the smaller margins which might be possible.

Motion calculated from CBCT data was applied retrospectively to plans, and the data used to calculate appropriate margins and also simulate online adaptive treatments with new margins, compared with conventional, using the ETHOS emulator environment. They found that an online daily adaptive process could enable a significant decrease in margin size (from the conventional 7–10 mm for these patients, down to 3 mm) and still provide excellent coverage and appropriate normal tissue sparing.

Calmels et al. (2023) also conducted an in-silico evaluation study using the Ethos autoplanning software retrospectively on a library of clinical anal (20), prostate (20) and rectal (20) patients. For each patient, the autoplanning software was used to generate three different IMRT plans (7, 9 and 12 field) and two VMAT plans (2 and 3 arc) for each patient, and they were compared using clinically relevant dose metrics with manually generated plans. All patients had standard preparation and simulation for treatment, so that all CT scans used were true to life for PCa treatment. All also had manual target and OAR contours delineated, which were then imported into the Ethos TPS for autoplanning. Plan quality and planning timings were recorded and evaluated (Calmels et al. 2023).

All plans were found to be acceptable and passed patient-specific QA tolerances, although the VMAT autoplans showed slightly reduced target homogeneity and slightly higher OAR doses than the manual plans; some IMRT autoplans showed improved conformity and homogeneity compared with VMAT manual plans. The median timescale for autoplan generation for the five plans across the 20 prostate patients was 39 mins. Within Calmels et al. (2023) paper, the two workflow patterns are illustrated and compared, with the total Ethos autoplanning process (for which the timescales were recorded) including provision for (i) selection of planning template (contours); (ii) import of contours; (iii) automatic plan generation; (iv) plan review and approval; and (v) automatic export for independent plan-specific QA checks. Optimum plan use (for large- and small-volume prostate patients) could be determined from the overall results, in order to maximise time-efficiency for possible online use.

Zwart et al. (2022) conducted a clinical feasibility study for the introduction of online CBCT-guided ART for PCa. Eleven PCa patients were treated with CBCT-driven online ART for a moderate hypofractionated regime (20 fractions). The on-treatment adaptive workflow (see Archambault et al. 2020 and details of that work in Chapter 10) involved (i) "influencer" review; (ii) target review; (iii) scheduled (recalculated) and newly adapted (reoptimised) plan generation and review; (iv) independent QA and (v) treatment delivery. The "influencers" are normal organ structures automatically detected by the system algorithms after CBCT image acquisition; they are structures closest to the target volume, which probably have the greatest impact on target shape and position and (post approval by the clinician) are used by the algorithm to propagate target structures (e.g., the CTV) from the scheduled planning CT onto the CBCT image to create a session model, ready for reoptimisation and adapted plan generation (Archambault et al. 2020; see also Chapter 10 on AI in this book). MR images used during scheduled pre-treatment simulation and planning are available at the treatment console to aid on-treatment influencer and target review during the process (Zwart et al. 2022).

Overall treatment times were recorded (as well as sectional process times) for the 11 patients, together with an evaluation of plan quality and reasons for adaptation, on the Ethos platform, an O-ring (Halcyon) style rotational delivery platform with improved acquisition speed and iterative kV-CBCT image reconstruction capabilities (Kirby and Calder 2019, chapter 10). Patients had treatment preparations (e.g., bladder and bowel preparation) and simulation (CT and MR scans) identical to non-adaptive PCa RT patients. On-treatment, two plans are presented on the CBCT anatomy – one recalculated from the original scheduled plan and one from the new, adapted and reoptimised plan. A second CBCT is acquired during the plan selection and QA process, and patient positioning is adjusted (using FMs), if necessary, for final set-up prior to treatment delivery. If the final set-up is deemed inappropriate, the adaptive process is restarted.

The full staff team at the treatment console consisted of two radiation therapists, a medical physicist, a clinical oncologist and a technical physician. The mean total treatment time was found to be 17.5 mins (10.8–28.8 mins), recorded as the time from the first CBCT acquisition to the end of treatment delivery. Broken down, the mean times for influencer generation and review, target generation and review, and plan generation and review were 6.5, 0.8, and 4.6 mins, respectively. The adapted plan was chosen for all fractions delivered, showing generally increased PTV coverage compared with the scheduled plan. In 19 out of 220 fractions (9%), the V60 for the bladder, rectum or both were above constraint for the adapted plan, with tenof the fractions which violated the constraint also being violated in the scheduled plan as well. For all fractions, the influencers required clinical edits, but for none of the fractions did the propagated targets need editing, nor were any set-up shifts required following the second CBCT scan.

Potential limitations highlighted included the lack of intrafractional motion monitoring. But even with the second CBCT scan, the motion was deemed acceptable, indicating the potential for reducing the CTV-PTV margins from the 7–8 mm currently used, from the non-adaptive treatment protocol but used for these patients. Overall, the clinical implementation of this CBCT-guided online adaptive protocol was deemed acceptable for a high volume of PCa treatments, with acceptable adaptive plans and a workflow manageable within an average 20-minute treatment slot.

McComas et al. (2023) performed an evaluation of costs and resource implications for CBCT-guided online ART for a number of pelvic disease sites, including the prostate. The online adaptive process map mirrors that used in Zwart et al. (2022) above for the Ethos system, including CBCT acquisition, online adaptation and a second CBCT before treatment delivery. The personnel involved were a radiation oncologist, physicist, dosimetrist, and three RTTs, all at the LINAC control desk. Time-driven activity-based costing analysis was used to evaluate the costs and resources associated with this type of adaptive treatment and assess whether the dosimetric benefit was worthwhile.

Results showed that a median extra time of 15.97 mins (interquartile range (IQR) 13.23–18.83) was required for each adaptive fraction, calculated to be approximately an extra \$100 per fraction. Dosimetric differences between scheduled versus adapted plan had a mean of approximately 16% (McComas et al. 2023). For treatments which already have good outcomes (PCa patients with current standards of IGRT), it is difficult to assess at this stage if the added resource requirement is cost effective; clear clinical results will help to add further information to the analysis.

Within the Ethos system, the main adaptive process is AI-driven and is published in Archambault et al. (2020). Details of the process can be found in Chapter 10, on AI, in this book.

Byrne et al. (2022) also used a study with data from 12 prostate patients (prostate alone, prostate bed and PLNs and prostate and PLNs treated on Halcyon) with the Ethos emulator to simulate Ethos online adaptive treatments; and compared the data with thse from six clinical prostate patients.

Bladder filling reduced slightly (from 500 ml to 400 ml) for the clinical cases because of the extended treatment times; for the retrospective 12 cases, all had 500 ml of water prior to simulation and treatment. IMRT (7, 9, or 12 field) techniques were used; VMAT was not, because of poorer plan quality observed previously (Sibolt et al. 2021; Calmels et al. 2023)

Clinical treatments required a staff complement of at least two RTTs and a physicist under the supervision of a prescribing radiation oncologist who, depending upon the progress of treatment, attended in-person or online. Post-adaptive CBCT was employed (as in Zwart et al. 2022), with assessment of intrafractional motion during the adaptive timeframe. If necessary, translational shifts were applied before treatment delivery. Timings were recorded from both the emulator and clinical treatments. For the retrospective study on the emulator, timing represented the simulated completion of CBCT acquisition to the end of plan acceptance. For clinical treatments, the time was recorded from the opening of a patient on Ethos to closing the patient at the end of the session.

In terms of study metrics, 11% of the AI-generated influencer contours required no changes, with 81% requiring only minor edits. Across all targets, 72% of cases required no editing, and the adaptive plan was the preference in 95% of fractions; meeting more planning objectives than the scheduled plan in 78% of fractions and being equivalent in 15% of fractions. In terms of timings, for intact prostate-only treatments, the retrospective, emulator data produced an average time of approximately 15 mins; within the clinical data set, the average timing was just under 34 mins, from opening the patient in Ethos to closing at the end of the session. For all sites on the emulator, the online adaptive, recontouring and replanning process would be possible within 20 mins; data which are different from those of Zwart et al. (2022), but the points of measurement may not be coincident.

With 92% of influencer contours requiring little or no edits and 91% of target contours the same, and with 95% of adaptive plans being selected for treatment, the data here are highly promising, showing the possibility of adaptive treatments with improved dosimetry, and a likely overall reduction in treatment time with increasing staff confidence and experience within the system.

The paper by De Roover et al. (2021) examined plan quality for the SBRT hypofractionated treatments for PCa with a simultaneous integrated boost (SIB) focal boost, of standard plans (C-arm LINACs) compared with Ethos autoplans. *In-silico* treatment planning study, based upon 20 cases of PCa with intermediate or high risk; previously treated as part of the FLAME (four patients) or hypo-FLAME (16 patients) trials (see Chapter 8; Draulans et al. 2020; De Cock et al. 2023).

Original planning for the trial was conducted for a TrueBeam STc C-arm LINAC with HD120 multi-leaf collimator (MLC) using 6MV flattening filter free (FFF) dual VMAT arcs. Manual planning was then conducted for the Halcyon fact-rotating O-ring system (Kirby and Calder 2019, chapter 10), which is the technology platform for the Ethos system. The Halcyon system plans were 6MV FFF dual and triple VMAT arcs. Finally autoplanning was conducted using the Ethos TPS, but only for triple VMAT arcs (because of machine limitations at the time with exceeding machine monitor unit tolerance limits per arc for the Halcyon system). Prescription was 35 Gy in five fractions to the prostate (30 Gy to the SVs) with an iso-toxic focal boost to the DIL (identified using mpMRI) of up to 50 Gy.

Results showed an enhanced target coverage for the Ethos autoplans compared with the C-arm LINAC plans, with increased SV PTV coverage and a reduced high-dose spillage to the bladder and urethra. However, there was a greater high-dose spillage to the rectum. Overall, the autoplans generated with Ethos for the fast-rotating O-ring LINAC system (Halcyon) were of similar plan quality to manual standard C-arm LINAC plans with high definition MLC and were physically deliverable (as evaluated by pre-treatment patient-specific QA using electronic portal imaging device (EPID). dosimetry).

In Moazzezi et al.'s (2021) *in-silico* study, 25 previously treated PCa patients were put through the Ethos emulator to assess autoplan quality for contours generated using the Ethos AI algorithms but without any editing. CTV was defined as the prostate and proximal SV volume; none had PLN involvement. Margins used were 3 mm posterior, 5 mm left/right and anterior and 7 mm superior/inferior. Adapted plans were calculated for ten fractions per patient utilising the Ethos auto-segmentation and autoplanning software without manual edits. Dose metrics from adapted plans were compared with unadapted ones to evaluate changes in CTV coverage and OAR doses.

Over 96% of fractions were deemed to require some form of editing following autosegmentation, although these were regarded as minor. Target coverage was found to improve for the adapted plans compared with non-adapted plans; and also reduced some OAR doses considered to exceed clinical thresholds. Adaptation with minor autosegmentation errors produced higher CTV doses for approximately 93% of fractions and improved the OAR sparing.

However, one patient was observed to have a major segmentation error which, if unnoticed and left un-edited, would have led to severe underdosage of the CTV. So, scrutiny of the autosegmentation process is still necessary, and not all patients may be suitable for plan adaptation.

Recently published is a general review (including prostate studies) of CBCT-based online ART – in essence, exploring the current trends of Ethos, including improved CBCT quality, clinical workflow, daily automated contouring and planning and motion management (Liu et al. 2023). The starting point is CBCT, whose quality itself has continued to improve through the use of iterative imaging methods (Kirby and Calder 2019, chapter 10) and the recent introduction of Hypersight with an extended field of view (FOV) (up to 70 cm), a further advance in reconstruction techniques and image quality, and faster acquisition speed of now less than 6 seconds. This has the potential for speeding up image acquisition for adaptive plans and also possibly improving the autocontour generation through the AI algorithm.

Five studies quoted within the review for prostate – data from Byrne et al. (2022), De Roover et al. (2021), Moazzezi et al. (2021), Morgan et al. (2023) and Zwart et al. (2022)– evaluated a total of approximately 900 fractions. Those evaluating ART procedural times reported averages between 10.7 and 19 mins.

Overall, the studies examined show that online CBCT-driven ART for PCa is feasible with Ethos. AI-driven auto-segmentation tools perform well with either no changes or minor edits, ensuring a procedural online adaptive time frame of 15–30 mins, based on the complexity of the anatomical site or treated volume for PCa. Dosimetric gains are observed, improving target coverage. Ongoing clinical studies will help define clinical gains regarding tumour control, morbidities, toxicities and patient-reported outcomes (Liu et al. 2023). Some studies show the potential for reducing staffing on set for adaptive treatment delivery, through in-person/online attendance (Byrne et al. 2022); potential staff changes with appropriate training and development in MRgART were put forward by Smith et al. 2023.

Work has been done to examine the clinical implementation of Ethos for the pelvic region (Sibolt et al., 2021). Autoplans for 39 pelvis cases, including prostate, and 100 online ART simulations were conducted with a pre-clinical release of Ethos. Evaluation was of plan quality compared with institutional reference plans; as well as AI-segmentation accuracy, online ART feasibility and integrated QA process.

More than 75% of AI-segmented contours required no or minor edits, with the associated adapted plan being deemed superior to the original clinical plan in 88% of cases. First clinical patients (not prostate) had a median timing of 17.6 min (from acceptance of acquired CBCT to start of treatment delivery); influencer segmentation was approximately 3–6 mins and review of propagated CTVs within an additional 1–3 mins. The remaining time was for plan generation, review, plan QA and occasional extra verification by CBCT imaging before treatment delivery.

Montalvo et al. (2022) published a case report for ART in radiation salvage of PCa; salvage for locally recurrent PCa after definitive radiotherapy is a challenging scenario, given risks to normal tissue under re-irradiation. This small, five-patient study had treatment on an adaptive online CBCT platform or on a 3-T-based MR LINAC. The added times for treatment with ART were, for the CBCT Ethos system, an added 10–15 min; for MRgART on an MR LINAC, an added 30–45 min. All patients completed planned treatments with excellent target coverage and acceptable acute toxicities, although follow-up time at publication was too short for subacute/late toxicity evaluation.

Similarly Ward et al. (2022) published early adopter experience with Ethos in the UK. The case was upper abdomen (lower oesophagus), with each adaptive fraction, from start to finish, taking less than 30 minutes. Like many of the other studies here, some editing of influencer structures was needed each day; automated (propagated) CTV and PTV contours did not require editing. In comparison, each day of adapted and scheduled plans, the adapted plan was chosen each time, demonstrating better target coverage with a more conformal dose distribution and fewer hotspots. OAR doses were found to be comparable or better each day with the adapted plan compared to the scheduled one.

Some studies in press at the moment are those of Byrne et al. (2023), and Waters et al. (2024). Byrne's work is an *in-silico* study examining 65 previous Ethos online ART patients. CTV was manually outlined on CBCT pairs, pre-adaptation and post-adaptation (the latter being a verification CBCT immediately before beam-on). They found that, if a shift is performed using the verification CBCT, a 4 mm margin on the prostate (and 5 mm on SVs) would be acceptable,with 95% CTV coverage in more than 90% of simulated fractions. When margins are applied to a simulated Ethos online ART workflow, CTV coverage was better than a standard IGRT workflow with 7 mm margins. If an online adaptive workflow is used without the associated verification imaging, larger margins may be required.

For Waters et al. (2024), they found that CT-based online ART improves target coverage and OAR avoidance for SBRT in PCa patients. This was a seven-patient study, treated with an SBRT UHF regime of 35 Gy in five fractions to the prostate and SVs (where six patients additionally had 25 Gy to the PLNs and five patients had a focal boost up to 40 Gy), using the CBCT-driven online adaptive process for all patients.

All seven patients completed treatment, with daily adaptation, resulting in a significant mean improvement to PTV coverage compared with scheduled plans. Mean rectal doses were significantly reduced, but there was a slight increase in bladder dose, although the bladder constraints were met for all fractions. Clinically, no patients experienced Grade 3, or worse, GI/GU toxicities within the median follow-up period of 9.5 months.

Finally, Price et al. (2023) published their experience as a single institute, having both CT-guided ART (CTgART) and MR-guided ART (MRgART) equipment. Their paper shares experiences for both Ethos and MRIdian, for a number of clinical sites, including the prostate. As of September 2022, 256 patients had been treated with adaptive intent: 186 with MRgART and 70 with CTgART.

From their institution, a comparison of treatment time slots for patients for adaptive prostate treatments showed an average of 70 mins for an MRgART procedure, compared with 45 mins for a CTgART procedure.

Final comments are needed concerning an MR-only pre-treatment workflow, one which is developing rapidly, especially for pelvis/prostate patients. The intention is clear: to remove CT from the planning process for the advantages that it brings on many counts, not least reducing concomitant dose and eliminating CT/MR fusion registration uncertainties (Kirby and Calder 2019, chapter 10). But to do so has many challenges, not least MR image distortion and the lack of electron density information within the MR image dataset. Nonetheless, many such challenges are being overcome, enabling a clinical implementation of MR-only workflow into radiotherapyT practice – for treatment on X-ray image-guidance-based RT equipment. For this to be possible (for standard IGRT and ART on these pieces of equipment), synthetic CT has been developed as the way forward to provide (a) electron density information for dose planning calculations and (b) CT scans and digitally reconstructed radiographs (DRRs) for reference images for on-treatment image guidance.

Although lack of room precludes a detailed account here of these advances for PCa patients, the reader is encouraged to consult the following papers for up-to-date research and shared experiences with synthetic CT: Johnstone et al. 2018, Bird et al. 2019, Speight et al. 2021, Gonzalez-Moya et al. 2021, Autret et al. 2023, and Wyatt et al. 2023a, 2023b.

5.4 MOTION MONITORING/MANAGEMENT

Over the past few years, results from clinical UHF trials for localised PCa patients have shown that these regimes are safe and noninferior to conventional fractionation, with follow-up data going beyond five years in many studies (see Chapters 8 and 9). The employment of these regimes is likely

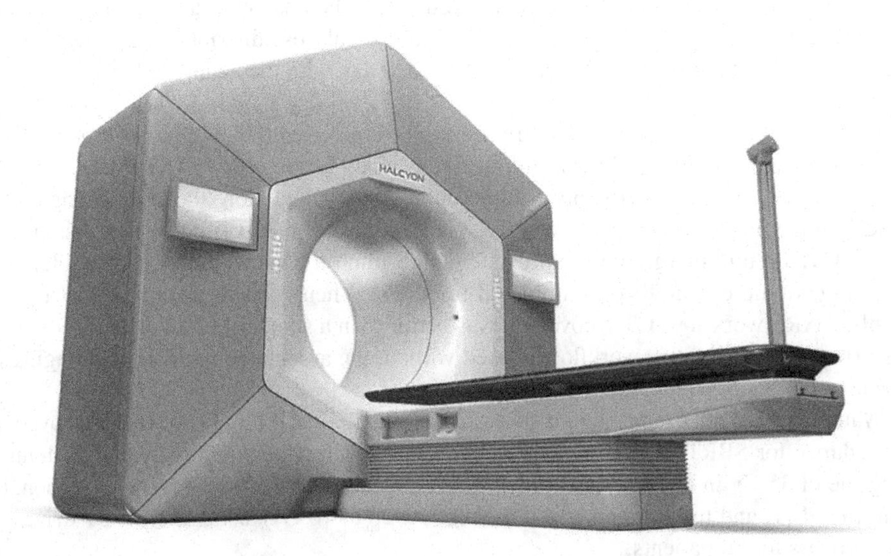

FIGURE 5.3 A photograph of the new Halcyon treatment delivery system; the platform on which the Ethos Adaptive Radiotherapy System is based. Image courtesy of Varian Medical Systems.

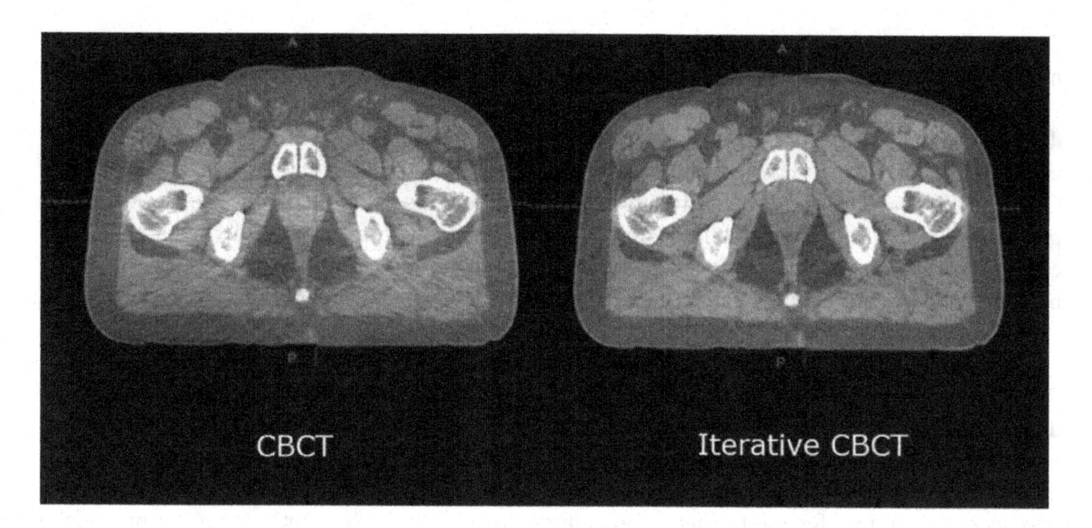

FIGURE 5.4 A sample image of the types of image quality improvements which are possible with iterative CBCT reconstruction compared with standard kV CBCT. This imaging is available and used for the Ethos Adaptive Radiotherapy System. Image courtesy of Varian Medical Systems.

to continue to increase and probably become a standard of care (Oehler et al. 2022; Loblaw 2020). But such treatment regimes can lead to prolonged treatment times (as can the ART process), compared with conventional fractionation methodologies (especially VMAT-delivered ones) and even moderately hypofractionated fractional doses. Times depend not only upon the fractional dose, but also on the treatment technology in use, the techniques applied and the beam modulation, e.g., standard C-arm LINACs with/without flattening filter (FF), Cyberknife, MR LINACs, online CTgART, Tomotherapy or VMAT relative to IMRT, etc. (Oehler et al. 2022).

Prostate motion while the patient is on the treatment couch is inevitable and non-periodic (Tudor et al. 2020), so the displacement can depend upon the duration of the treatment delivery. Evidence suggests that shortening the time can reduce the risk and overall dosimetric effects of intrafractional motion, with data suggesting that prolonging treatment duration (say, from 5 to 10 min) can result in motion greater than 3 mm, thereby challenging the margins set during planning and risking underdosing the target tissues and overdosing nearby OARs (Ghadjar et al. 2019; Bertholet et al. 2019). This can naturally be ameliorated with the use of enhanced margins, but with the inevitable consequence of greater toxicity, especially acute for UHF treatment techniques with high conformity, which are ideal techniques for the prostate to achieve a high therapeutic index.

This section shares some of the techniques used with PCa for monitoring the intrafractional motion which may be present during treatment and the techniques used to compensate for it through beam interruptions (and often reimaging/reset-up) or tracking/adapting in real time. Methods include the use of FMs and X-ray-based in-room and gantry-mounted technology, non-ionising methods (see Chapter 4; Kirby and Calder 2019, chapters 7 and 8), such as ultrasound, electromagnetic (EM) transponders and the latest MR LINAC technology; the non-ionising ones naturally having the benefit of zero concomitant dose burden.

For the prostate, two main types of intrafractional motion which have been identified are (i) non-resolving slow drift, mainly in the anterior/posterior (AP) direction, due to rectal filling and (ii) sudden, transient shifts, mostly in the AP and superior/inferior (SI) directions because of bowel peristalsis (Tudor et al. 2020). Pelvic muscle clenching has also been known to have an effect, with longer treatment times (which may be the case with SBRT and/or adaptive RT processes (both MRg and CTg)) naturally increasing the risk of the effects of intrafractional motion (Ghadjar et al. 2019). Potentially, this can make the motion significant dosimetrically for both target and nearby OARs. Immobilisation approaches using rectal balloons may limit the intrafractional motion (see

Chapter 2), especially for treatments beyond about 2.5 mins; recto-prostatic spacers, however, do not appear to significantly influence the movement during treatment (Ghadjar et al. 2019).

Within the latest national guidance on assessing geometric uncertainties, using predominantly daily online image guidance and calculating CTV-to-PTV margins accordingly (Tudor et al. 2020), tracking options such as US and implanted/inserted EM transponders (see Chapter 4; Kirby and Calder 2019, chapters 7 and 8) can be used to monitor and also stop treatment if motion exceeds pre-defined tolerances. The magnitude and standard deviations of measurements of such motions can be used in appropriate margin calculations. If intrafractional motion imaging/monitoring can't be conducted during actual treatment delivery, assessment of the motion can be made through volumetric imaging conducted before and after the treated fraction, and these data used to inform systematic and random error components and therefore the margin calculation (Tudor et al. 2020; RCR 2021). The overall residual margin will depend upon whether imaging/monitoring can be conducted throughout treatment delivery (with the potential to correct and make adjustments accordingly) or whether on-treatment imaging only takes place before delivery (with a post-treatment image taken to quantify and inform after the event). Margin calculations should take these strategies into account. So, too, for the timing of procedures (especially with adaptive ones, as mentioned in Sections 5.2 and 5.3), because intrafraction motion changes and increases with time while the patient is on the couch (Ghadjar et al. 2019; Tudor et al. 2020, pp. 25–26).

One of the most complete reviews of methods for intrafractional motion monitoring (across many different sites and with a wide range of technologies) is given by Bertholet et al. (2019), for monitoring both external surface and internal anatomy and using ionising and non-ionising radiations for the latter. Real-time monitoring is defined as measurements made, analysed and processed of target positioning with information from the acquired image (or device) within about 0.5 seconds for respiratory motion. For slower, non-periodic motions (like that for the prostate), the time delay can be longer. "Online", like on-treatment verification imaging in standard image guidance, refers to monitoring performed while the patient is on the treatment couch.

Most commonly for the prostate, internal monitoring is achieved using implanted FMs or EM transponders, where the markers act as surrogates for the prostate position because of poor soft tissue contrast with X-ray imaging. The advent of MR imaging on MR LINACs (with the potential for real-time cross-sectional imaging during treatment delivery) will naturally change the landscape in terms of obtaining much better contrast images and thereby not requiring invasive FMs as a prostate surrogate.

Conventional kV X-ray on-treatment imaging for real-time intrafractional monitoring/tracking has involved the use of orthogonal, in-room imaging systems, such as those which are part of the Cyberknife treatment system (Kirby and Calder 2019, chapter 10) and also stereoscopic in-room imaging used on standard C-arm LINACs (Kirby and Calder 2019, chapter 10). The Cyberknife system has an accompanying array of sophisticated image detection and monitoring algorithms which can be either FM-based (as for the use with PCa patients) or non-fiducial, anatomically based. Recent studies (Oehler et al. 2022) have investigated the use of Cyberknife for intrafraction prostate motion management for UHF PCa treatments. Making use of the monitoring data obtained on Cyberknife orthogonal kV in-room imaging, optimum imaging frequencies were determined, with the recommendation that intrafractional motion management (monitoring and correction) is recommended for UHF regimes when treatments exceed 2.5 mins. Imaging frequencies within 50 seconds allowed PTV safety margins for intrafraction uncertainties of 0.5–1 mm.

Use can be made of the on-board X-ray imaging available on most C-arm LINACs. The most noted examples for prostate include the kilovoltage intrafraction monitoring (KIM) system which produces real-time, 3D results on a standard C-arm LINAC (Keall et al. 2016), together with gating (e.g., through a 3-mm/5-second action threshold) where the beam is paused during treatment, or even in conjunction with treatment adaptation by tracking movements of the MLC leaves in real time (Keall et al. 2018, 2019).

More recent examples include Kuo et al. (2023), where kV intrafractional imaging is developed on a standard C-arm LINAC for use during VMAT treatments of PCa patients. Results show that the 2D kV method is feasible with a proper intervention scheme for UHF treatments. Prostate motion is identified over time intrafractionally, so, ideally, the elapsed time between the set-up images and start of treatment delivery should be minimised, X-ray imaging frequency optimised and overall treatment time shortened in order to minimise interruptions to treatment delivery because of intrafractional motion exceeding thresholds.

Further examples include combining MV and kV imaging during UHF VMAT treatments on standard C-arm LINACs (Gorovets et al. 2020). Implementation is feasible with results demonstrating that most movement is within 3 mm, but some patients have larger movements not accounted for by the applied PTV margin, movement which would have gone undetected/uncorrected without intrafraction motion management. Median numbers of interruptions for repositioning amounted to one per fraction, but median overall treatment time was still maintained to about 8 minutes. Clinical results for toxicity and control were acceptable for 151 patients treated and with a follow-up of greater than three months (Gorovets et al. 2020).

Maas et al. 2023 have reported on the use of triggered kV imaging for monitoring intrafraction motion for their VMAT-based UHF regime (36.25 Gy in five fractions) with an SIB to an MRI-defined focal lesion. Arumugam et al. (Arumugam 2016, Arumugam, Young, et al. 2023) demonstrated that data from their SeedTracker monitoring system would suggest that, if real-time monitoring and positioning were applied, then margins could be reduced to between 2 and 3 mm, whereas, for their patients and data, the margins would need to be 3–4 mm to maintain coverage.

MV imaging systems can also be used on their own, during treatment delivery, with the advantage of not adding to the patient's concomitant dose burden. Often described as Beam's Eye View (BEV) imaging, motion detection is 2D and is more difficult because of lower image contrast and the modulated fields, but can still yield useful motion information in the most important direction, namely perpendicular to the beam direction. Recent developments seek to use AI and machine learning to further improve further MV-based real-time imaging for PCa patients (Chrystall et al. 2023)

Dosimetrically, intrafractional motion has been shown to have a significant effect if it is not corrected for. In the dosimetric comparison by Colvill and colleagues (2016) between adaptive and non-adaptive tracking systems, scenarios were undertaken on multiple RT platforms (which included robotic, gimballed, MLC, and couch tracking systems). Data from such platforms of real-time motion traces for prostate patients were used to formulate and deliver plans to a moving dose-meter, with and without motion correction. For non-adaptive systems, the failure rate was reported as about 17%, compared with about 1.5% for adapted systems, showing that all the systems tested accounted for realistic tumour motion, performing to a very high standard.

Calypso and Raypilot/Hypocath EM transponder-based ultrasound systems have been shown to be effective in monitoring and correcting for intrafractional motion for PCa patients. For information on these systems, please see Chapter 4.

5.5 OTHER TECHNOLOGIES

The reader is encouraged to familiarise themselves with the IGRT capabilities in general for these technologies in Kirby and Calder 2019, chapter 10.

5.5.1 TOMOTHERAPY

Volumetric MV X-ray CT scans can be acquired before, during and after treatment in the integrated Tomotherapy system. By utilising the treatment beam for imaging, reconstruction of the delivered dose is possible in addition to its use for IGRT position verification prior to and during treatment.

However, MV CToften has a higher concomitant imaging dose and lower soft-tissue contrast than kV CBCT but fewer metal artefacts (Zou et al. 2018).

The latest system, known commercially as Radixact, now employs kVCT imaging with a separate kV X-ray source and imaging panel. This enables imaging in a helical fashion, very similar (and with similar image quality) to the standard helical fanbeam kV CT used on CT scanners for treatment planning and which is the source of reference image data for IGRT purposes.

Tegtmeier et al. (2022) have characterised the new imaging capability for both image-guided radiotherapy and ART, compared with Fanbeam MVCT, kV-CBCT and standard kVCT. Noise was considerably reduced and similar to that of standard kVCT and slight cupping/capping artefacts were noticeable compared to kVCT. Low-contrast object visibility and contrast-to-noise ratio (CNR) were again similar to kVCT and better than MVCT and kV-CBCT, while similar CT number linearity was recorded.

Initial indications show excellent promise in improved image quality over previous Tomotherapy MV-based imaging systems and superior to some kV-CBCT systems, making it adequate for image guidance and sufficient for structure delineation for ART purposes, with a similar concomitant dose burden (measured using CT dose index (CTDI)) to that of planning kVCT scans (Tegtmeier et al. 2022).

Another innovation in the Radixact system is now the real-time motion tracking with jaws and MLC during helical Tomotherapy delivery. This has been tested for lung and prostate patients by Chen et al. (2021). Known as Synchrony (originally developed for Cyberknife and the tracking of lung tumours (Kirby and Calder 2019, chapter 10), it has now been developed for the Radixact helical treatment delivery system.

For the prostate patients, four FMs were implanted under ultrasound (US) guidance, as for standard PCa treatments. Plans were created for treatment either with 70 Gy over 28 fractions (four patients) or 75.6 Gy over 42 fractions (one patient) for both Synchrony and non-Synchrony delivery, the latter acting as a backup in case of breakdown. Beam-on time for treatment ranged from approximately 150 seconds to 370 seconds (median 315 seconds) (Chen et al. 2021).

Like the Synchrony system used on Cyberknife, the technique combines X-ray-based imaging of internal anatomy (either through a surrogate like FMs or with feature detection of internal anatomy as used for markerless tracking in lung tumours) with externally mounted cameras within the treatment room to capture the motion of a set of light-emitting diodes (LEDs) placed on the patient's skin and the treatment couch. By tracking the LEDs while simultaneously acquiring X-ray-based images to detect tumour location, a model is developed, correlating internal and external motion (Chen et al. 2021).

During treatment delivery, tumour motion is then predicted from the model, updated from newly acquired X-ray images with motion compensated for using dynamic MLC shifts and jaw sweeping. Variables for fine-tuning the system and success of the model include imaging quality and angles, LED placement, FM selection and segmentation/thresholding. Manual or automatic beam-off can be initiated if movement exceeds a set threshold, within a certain timeframe, or if the FMs within the target move outside the tracking range (Chen et al. 2021).

Results showed that patient selection was an issue, as was optimal patient/LED placement, appropriate setting for tracking parameters and the use of treatment simulation. The latter point used a simulation tool to help guide the user as to which patients were or were not suitable for Synchrony treatment. The average model confidence for prostate (fiducial tracking) patients was between 0.96 and 1.00, and the overall percentage of beam-on radiographic imaging was between approximately 85% and 95% (median 89%). The concomitant dose (extra imaging dose) was estimated to be about 0.3% of the planned dose (approximately 0.21 Gy for the 70 Gy prescription patients). Intrafraction motion was estimated to be typically 2–5 mm within about 6 minutes of treatment, with some movements recorded of up to 8 mm. All target tracking metrics were within specification, indicating a very good agreement between existing and updated tracking models. Overall, the clinical motion tracking and correction system was found to work well within specification, and patient workflow and process characteristics were developed to inform refinement and further clinical implementation for selected prostate patients.

Mannabe et al. (2021) report on a planning comparison study, where 20 patients were investigated, for which two plans were created (simulating the Synchrony system) for each; one for Tomotherapy and one for Cyberknife, all with a dose of 36.25 Gy in five fractions and uniform margins of 2 mm around the CTV to create PTVs. Results showed that all plans achieved the desired target objectives and OAR dose constraints; Tomotherapy plans were deemed superior to Cyberknife for the rectum V80%, but slightly inferior for PTV-V110% and V80% for the bladder. Simulated treatment times resulted in a median time of 8.6 mins for Tomotherapy (range approximately 7–11 mins) and 16.0 mins for Cyberknife (range approximately 13–20 mins). Overall conclusions were that the data suggested that Tomotherapy plans with the Synchrony system would have reasonable potential for SBRT-style treatments for localised PCa patients with hydrogel spacers.

In Lu et al.'s (2023) *in-silico* study, the simulated plans for patients treated with Tomotherapy and the Synchrony motion-tracking system were examined. Ten retrospective PCa patients with actual prostate intrafractional motion curves (obtained through the Calypso EM transponder system (see Chapter 4 and Kirby and Calder 2019, chapters 7 and 8) were replanned on Tomotherapy using an SBRT hypofractionated prescription of 35 Gy in five fractions, in combination with different jaw settings (1 cm static, 2.5 cm static and 2.5 cm dynamic jaws). Two compensation methods were evaluated: one with a jaw and MLC adjustment every 0.1 seconds and one with a realistic compensation from kV imaging intervals of every 6 seconds. In-house 4D dose calculation software was used to evaluate the resultant dose distributions for (a) the original (motion-free) plan, (b) a motion-contaminated plan and (c) the two compensation strategies mentioned. For one patient, who had high motion amplitude data (> 5 mm), various imaging frequencies were also investigated (Lu et al. 2023).

Without motion compensation, the PTV coverage decreased significantly and bladder/rectum OAR doses increased. Motion compensation improved PTV coverage when combined with the 2.5-cm jaw, though being slightly less effective for the 1-cm jaw. The compensation also restored OAR sparing to the levels achieved in the original motion-free plans. The effectiveness of motion compensation was found to be dependent upon imaging rate. A typical rate of two images per gantry rotation (one rotation taking approximately 12 seconds for the Radixact Synchrony system) effectively reduced the motion-induced uncertainties, although for larger motions and a 1-cm jaw, an increase in imaging frequency would be better (Lu et al. 2023).

Gallio et al. (2023) undertook a planning comparison of Radixact and VMAT plans for different anatomical sites, including the prostate. Sixty patients were investigated with PTV coverage, OAR dose constraints and other dosimetric indices compared, together with delivery assessment using gamma evaluation during QA, beam-on time comparison and clinical judgement evaluation of the plans.

For prostate plans, gamma index values were highest for the Radixact system compared to the Versa HD VMAT plans; PTV doses were similar with equivalent conformity and homogeneity indices. Median values of V37Gy for the bladder and V28Gy for the rectum were both lower for the Radixact plans than for Versa HD plans. Clinical scoring showed that both systems produced similar plans, with median beam-on times being 517 seconds (456–706 seconds) for Radixact plans compared with 169 seconds (163–218 seconds) for Versa HD VMAT plans (Gallio et al. 2023).

5.5.2 Cyberknife

For Cyberknife, the use of FMs is needed for PCa patient target localisation and monitoring through the use of repeat intrafractional imaging using in-room kV X-ray systems (Kirby and Calder 2019, chapters 7, 8 and 10). Imaging frequency needs to be kept low to reduce the concomitant dose burden, but needs to be high enough to detect significant prostate deviations for either beam interruptions or, ideally, because of the design of Cyberknife technology, tracking the movement and thereby always maintaining the CTV within the desired dose volume (Oehler et al. 2022). Reports suggest imaging intervals of 40–180 seconds or recommendations dependent upon margin magnitudes, e.g., up to 240 seconds to maintain movement within 3 mm. But no robust guidelines exist.

This study (Oehler et al. 2022) was devised to address some of these issues and give guidance for the appropriate compensation mode – compensation with PTV margins or intrafractional motion tracking. Nine patients with localised PCa were examined, treated with UHF SBRT using Cyberknife. In-room kV stereoscopic imaging was used with an average interval of 19–92 seconds. The statistical distribution of motion was modelled and used to calculate PTV margins needed for imaging intervals of 0–300 seconds for a particular % probability of coverage (e.g., 98%) in a certain percentage of patients (e.g., 99%).

Results showed a need for PTV margins ranging from approximately 0.6 mm left/right (LR), 0.9 mm (anterior/posterior (AP) and 1.1 mm superior/inferior (SI) for a 40-second imaging interval, and up to 2.6 mm (LR), 5.3 mm (AP) and 5.6 mm (S.I) for a 300-second imaging interval to achieve a 98% probability of coverage in 99% of patients. This would suggest that, for treatment times less than 150 seconds, intrafractional motion is low and could reasonably be compensated for with an additional margin of 2–3 mm. Above these times, which may be the case for some UHF techniques and delivery technologies, motion monitoring and correction is likely needed. Previous work (Lovelock et al. 2015) using EM transponders showed that margins needed to increase by approximately 2 mm for every 5 mins of treatment delivery; data here would suggest similar or larger compensation may be required (Oehler et al 2022).

In a recent study published by Rose et al. (2023), the intrafraction motion present during Cyberknife-based SBRT for PCa patients was examined, with a view to optimising imaging frequency. Intrafraction displacement data were obtained for 331 patients, with a total of nearly 85,000 images over 1,635 fractions of treatment analysed. Results showed that FM distance movements for consecutive image pairs ranged from < 2 mm in 92.4% of instances to < 10 mm in 97.7% of instances. Treatment times per fraction were a mean of 35 mins (minimum 7 mins, maximum 1 hr 37 mins).

Worst-case displacement data would suggest:

- Mean percent of time CTV within 3 mm – 96%, 96%, 96% (SI, AP, LR) for 15-second imaging
- Mean percent of time CTV within 3 mm – 94%, 94%, 94% (SI, AP, LR) for 30-second imaging
- Mean percent of time CTV within 5 mm – 96%, 95%, 96% (SI, AP, LR) for 45-second imaging
- Mean percent of time CTV within 5 mm – 95%, 94%, 95% (SI, AP, LR) for 60-second imaging
- Mean percent of time CTV within 5 mm – 94%, 93%, 94% (SI, AP, LR) for 90-second imaging

As part of a multicentre, phase 2 trial for UHF SBRT boost for intermediate-risk PCa patients, Pasquier et al. (2023) published results from their study to assess late toxicities and relapse-free survival. For the trial, the seventy-six patients enrolled were treated between 2010 and 2013, with 60 (79%) using Cyberknife and 16 (21%) using linear accelerators. Doses of 46 Gy were applied in 23 fractions first (with either 3D-CRT or IMRT) and then a boost of 18 Gy in three fractions was delivered. CTV1 was defined as whole prostate plus the proximal half of SVs for the first part of treatment; boost CTV2 was the prostate alone, defined using MR and CT registration based on FMs. Margins were 1 cm around CTV1, 5 mm posteriorly; PTV2 was defined from a 5 mm expansion around CTV2.

With a median follow-up of 88 months (81–99 months), Grade 2 GU toxicities at 120 months were 1.4%, with GI toxicities at 11.0%. Overall and relapse-free survival were 89.1% and 76.9%, respectively, at 8 years of follow-up. Overall, for this treatment involving a hypofractionated boost after conventional fractionation, low toxicity was found with good survival rates at a long-term follow-up, suggesting a safe method using Cyberknife and standard LINAC delivery (Pasquier et al. 2023).

5.5.3 Biology-guided Radiotherapy (BgRT)

For an introduction to this new technology, see Kirby and Calder 2019, chapter 10. Since that publication, the image guidance capabilities for the treatment unit have been developed further and a number of papers have been published, testing different aspects of the image guidance systems. Oderinde et al. (2021) published the latest technical details of the BgRT system (known commercially as RefleXion X1). It now has several IG systems on-board, including MVCT, a 16-slice kvCT imaging system and positron emission tomography (PET), the latter enabling both functional imaging and biology guidance for treatment delivery – with the tumour's emitted PET profile (biological signature) acting as a fiducial marker, increasing the confidence in lesion localisation during actual treatment delivery (Oderinde et al. 2021; Kirby and Calder 2019, chapter 10).

The BgRT tomographic delivery algorithm requires no surrogate motion sensors, predictive motion models or breath-hold coaching, and detection is with sub-second latency. PET imaging is used to (i) collect PET data for treatment planning, (ii) acquire a pre-treatment PET scan and (iii) actively guide the therapeutic beam during treatment.

Oderinde et al. (2022) have gone on to test the kVCT scanning capability on the RefleXion X1 for post-prostatectomy patients treated with 44 Gy in 22 fractions to the prostate bed and PLNs. Five scans were acquired throughout the treatment course, and the target and OAR structures were recontoured onto the CT scans. IMRT treatment plans were recreated on each daily kVCT scan and doses were recalculated for comparison with the original planning CT. Interfractional changes were analysed with planning dose metrics as well as by examining correlations between bladder/rectum volumes and mean doses.

Median dose changes were found to be approximately 0.5 Gy for the bladder and 0.6 Gy for the rectum, together with other metrics, showing that, with adaptive planning on the daily kVCT images, interfraction variation can be minimised. A correlation is still observed between bladder volume and dose. The study showed that the kVCT images were of sufficient quality for target and OAR delineation and dose calculation, paving the way for further adaptive planning during treatment (Oderinde 2022).

Gaudreault et al. (2022, 2023) have published work on some of the unique features of the RefleXion X1 BgRT platform, namely BgRT with the use of ^{177}lutetium prostate-specific membrane antigen (PSMA) treatment for metastatic PCa. Their work demonstrated the flexibility of the platform to combine functional imaging with biology to guide the radiotherapy delivery, alongside novel theragnostic treatments, with the possibility of delivering biology-guided treatment to PSMA-negative/FDG-only avid tumour deposits, identified because of the different pathways of the biochemical analogues, namely fluorodeoxyglucose (FDG) and gallium-prostate specific membrane antigen (Ga-PSMA). This feasibility study was found to show promise in combining the treatments for patients with PSMA/FDG discordant metastases.

Ketcherside et al. (2023) undertook a study to examine the potential of this new technology; by assessing its different imaging protocols and image quality for the purpose of radiomics analysis, for which repeatability and reproducibility of the imaging subsystems are extremely important. For PCa, radiomics is the extraction and analysis of quantitative images that can be used to develop descriptive and predictive cancer models, by combining image features and phenotypes with gene and protein signatures (Ferro et al. 2022). Radiomic feature extraction (from multiple imaging modalities, like those on RefleXion X1 in BgRT) and analysis can aid detection, risk stratification and treatment for PCa patients. Combined with machine learning techniques, comes the possibility of differentiating between, for example, low- and high-grade PCa, improving tumour description and risk assessment, and aiding in treatment planning (Ferro et al. 2022).

Ketcherside et al. (2023) tested the kVCT imaging platform using a specific Radiomics phantom, composed of different materials that mimic the radiomics features of different tumours. For this particular type of Radiomics phantom, the kVCT imaging system showed excellent reproducibility and stability over time, being suitable, therefore, for quantitative imaging for Radiomics purposes of texture and feature extraction in tumours like PCa.

5.6 TRAINING/WORKFORCE

Many of the papers above (especially in the MR and CT sections, Sections 5.2 and 5.3, respectively) indicate a greater resource requirement (in terms of staffing complements) that is needed at the radiotherapy treatment console for online ART, whether it be MR or CT guided.

For the Ethos CtgART systems, groups have mentioned staffing complements such as:

- Zwart et al. (2022), where online ART requires two therapeutic radiographers (RTTs), a medical physicist, a clinical oncologist and a technical physician.
- McComas et al. (2023), where the personnel involved are a radiation oncologist, a physicist, a dosimetrist and three RTTs, all at the LINAC control desk.
- Byrne et al. (2022), where the staff complement comprised at least two RTTs and a physicist under the supervision of the prescribing radiation oncologist who, depending upon the progress of treatment, attended in person or online.

Byrne et al. (2022) show that staff complements can be changed, and, most likely, roles and responsibilities changed accordingly for what is a resource-hungry process, in having experienced staff busy with online tasks in one place for the whole treatment day, if one looks eventually to a normal LINAC workload in a busy department.

Since most treatment units are run by a complement of RTTs, usually leading on all tasks (including online and offline IGRT), who are needed for the patient's treatments, many therapeutic radiographer groups in the UK have been active in examining how staff complements can change – safely and effectively – for this complex process that is online ART.

Hales et al. (2020) published work on their experience and how the process could be made more radiographer-led. They note, as early adopters of the Elekta Unity MR LINAC, the daily multidisciplinary presence they set up for delivery on this new technology; and have been investigating ways in which it could be made more standardised and of greater similarity to normal IGRT practice in the UK, i.e., radiographer led. Work, therefore, was undertaken to identify the knowledge, skills and competence required to achieve this, examining each step of the MR-guided ART (MRgART) workflow.

Through an informal survey of all staff groups involved in their initial clinical model on the MR LINAC and undertaking a needs assessment to identify additional and enhanced skills needed, they performed a critical evaluation and encouraged open discussion from all disciplines to iteratively develop a new model. They began with the simplest of online MRgART strategies (virtual couch shift and online replanning); known as Adapt-to-Position (ATP) (Tocco et al. 2020; Winkel et al. 2019; Poon et al. 2022a; Dassen et al. 2023; Lawes et al. 2022; see Section 5.2 above).

Steps to implementation included (i) critically evaluating the ATP workflow for prostate, skills required, etc., after treatment of their first two patients; (ii) recognising and identifying the staff groups required to have the appropriate MR LINAC skills at each step of the ATP workflow; (iii) discussing and agreeing tolerances and action levels for intervention, aided by a traffic light system; (iv) delegation of clinical oncologist responsibilities for certain steps, if needed, to specific MR LINAC staff with agreed competencies and establishment of an on-call system; and (v) developing a specific protocol for a "clinician-lite" prostate ATP workflow which is agreed universally.

Timings of patient workflows, before and after modification, were recorded for two patients in the "fully staffed" framework and ten patients for the "clinician-lite" one. Various teaching and training methods were used for the development of knowledge and skills for the teams, including, for example, vendor training, external and internal courses, tutorials, workshops and self-directed learning (see Chapter 11). Wide-ranging MR-focussed subject matter, from basic MR safety and physics through to MR image acquisition and interpretation, especially with regard to anatomy on MR images, treatment plan evaluation, etc., was devised and undertaken.

As a result, staff complements for the prostate ATP workflow have changed over the course of development and implementation to the "clinician-lite" environment, and further streamlining is likely to continue. For the two prostate patients treated (40 fractions in total), who had a full staff complement (of clinical oncologist, two physicists/dosimetrists and two therapeutic radiographers), the mean in-room treatment time was just over 36 mins, and the clinical oncologist was present and involved for 100% of the time. For the ten patients treated using the "clinician-lite" workflow (200 fractions in total and a staff complement of one physicist/dosimetrist and two therapeutic radiographers), the mean in-room treatment time was just over 31 mins, and the clinical oncologist was present and involved for just 1.3% of the time.

The reduced staffing model has therefore been successfully introduced for the more simple ATP ART model on the MR LINA for radical prostate patients, demonstrating an improvement in efficiency and enhancement of the therapeutic radiographers' profile, approaching a model which is more like conventional IGRT on standard LINACs.

McNair et al. (2020) examined staffing complements through an international survey conducted on current practice with online adaptive RT, using MR guidance, xamining principally the implementation phase of a clinical service and the professional roles and responsibilities utilised in the workflow, together with projections for future service development.

A 38-question survey (circulated electronically through radiotherapy networks) was used, covering areas such as (i) current practice; (ii) professional responsibilities; (iii) benefits and barriers; and (iv) decision-making responsibilities. The ART process considered was one where target/OAR tissues are contoured and replanning was conducted online.

A total of 19 international responses were received from across Europe (11), the USA (4), Canada (2), Australia (1) and Hong Kong (1). Most centres had been operating for 1 or 2 years, with the first having started ART five years beforehand. Most had started with prostate and oligomets patients (with oligometastatic cancer) and most had started with conventional staffing roles but were moving towards a greater responsibility for radiographers in contouring and treatment planning. The three most important criteria identified by clinical oncologists for moving into ART treatments were (i) overall gross anatomy changes for target and OARs, (ii) inadequate coverage of the PTV, and (iii) high-dose volume being too close to OARs. There were no clear guidelines for the degree of improvements needed for target and OAR volume doses.

The results showed that successful implementation has taken place internationally, beginning with workflows using traditional staff roles, but larger complements on the treatment units during ART procedures. From the responses, the median numbers required for treatment (with associated range) were: (a) physicist: 1 (0–5), (b) dosimetrist: 0 (0–4), (c) radiographer: 2 (1–5), and (d) clinical oncologist: 1 (0–5). The data are confounded slightly when, for some countries (like the Netherlands), many radiographers also undertake dosimetrist tasks, with a similar skill mix being encountered in Australia and New Zealand. In terms of online recontouring during the adaptive process for PCa patients, nearly 50% of centres undertook this for OARS for every fraction; 45% of centres were doing so for target tissues for every fraction. Streamlining for some roles (such as the clinical oncologists being on the MR LINAC) could come from, say, securely designed automated procedures checking for anomalies and requiring direct clinical input only when constraints/objectives are breached beyond well-defined tolerances. It can be noted that, in the CTgART environments too, staff complements are being examined and changed from a 100% on-site presence for all staff disciplines (Byrne et al. 2022).

Faster software and greater staff resourcing seemed to be the largest barriers to implementation of ART. Although most centres started with conventional staffing complements for the workflows, this is beginning to change with a fresh consideration of professional responsibilities and the implementation of new training programmes.

As part of that consideration, McNair and colleagues (2021) have continued their work in examining the place of traditional roles and responsibilities on the MR LINAC and exploring new ways of working towards a radiographer-led online MRgART process. For this study, radiographers, clinicians and physicists from centres with MRgART experience were surveyed through interviews of a focus group (of 6-8 individuals), in order to allow expansion of each other's responses, to obtain several perspectives and derive a consensus. A semi-structured interview approach was taken using two facilitators.

Questions and discussions examined professional background and ART experience; the positive and less positive aspects of their roles; their training undertaken, methods used, effectiveness and things missing. They explored future learning and training needs, and methods of delivery, communication, and experience and understanding of ART workflows, etc.

From the responses and framework analysis, three key themes were derived: (a) Current MRgRT practice with roles of staff across different workflows and impacts on the team; (b) Training, with methods, timing and process; and (c) Future practice, with barriers and drivers for developing and implementing MRgRT.

Overall, the development of a radiographer-led MRgRT practice was seen as an exciting opportunity for role development. Ideally, this would be undertaken through a national training framework created between all stakeholders and professions, which could provide a solid consistency in required skills and knowledge. An MDT culture, with effective communication encouraging and promoting shared learning, was seen as vital for the development and transfer of staff roles for this online MRgRT ART process.

Their work led to formal research, investigating and testing the effectiveness of new training and development for the important work area of contouring within an online ART environment. Smith et al. (2023) have recently published their results to evaluate radiographer-based contouring for online MRgART on PCa patients. Following robust and specifically designed training programmes, an online interobserver variability study was conducted, examining 117 images from six patients on the MR LINAC, contoured online by either a radiographer or a clinician, with the same images contoured offline by the alternate profession. Metrics such as Dice Similarity Coefficient (DSC), mean distance to agreement (MDA), Hausdorff distance (HD), volume metrics and further measures on optimised plans (from radiographer online contours) were used to analyse and compare the contours. Following clinical implementation of radiographer contouring, target volume comparison and dose analysis were then performed on 20 contours from five patients.

No significant difference was found in volume sizes between the two groups; aided by the similarity metrics in a quantitative fashion. Clinician review (blind testing) of fractions using trial QA metrics deemed all plans to be acceptable. Overall, radiographer and clinician prostate and SV contouring on MRI within an online ART workflow have been found to be comparable, producing clinically acceptable plans, and indicating that clinical implementation can be introduced effectively with appropriate training and robust evaluation processes. Experience here also shows that a DSC threshold for target structures could be implemented to streamline future training.

REFERENCES

Acharya, S., Fischer-Valuck, B., Kashani, R., et al. 2016. Online magnetic resonance image guided adaptive radiation therapy: First clinical applications. *International Journal of Radiation Oncology, Biology, Physics*. 94(2): 394–403.

Antico, M., Prinsen, P., Cellini, F., et al. 2019. Real-time adaptive planning method for radiotherapy treatment delivery for prostate cancer patients, based on a library of plans accounting for possible anatomy configuration changes. *Plos One* 14(2): e0213002.

Archambault, Y., Boylan, C., Bullock, D., et al. 2020. Making on-line adaptive radiotherapy possible using artificial intelligence and machine learning for efficient daily re-planning. *Medical Physics International Journal*. 8(2): 77–86.

Autret, D., Guillerminet, C., Roussei, A., et al. 2023. Comparison of four synthetic CT generators for brain and prostate MRonly workflow in radiotherapy. *Radiation Oncology*. 18: 146.

Bertholet, J., Knopf, A., Eiben, B., et al. 2019. Real-time intrafraction motion monitoring in external beam radiotherapy. *Physics in Medicine and Biology*. 64: 15TR01.

Bird, D., Henry, A., Sebag-Montefiore, D., et al. 2019. A systematic review of the clinical implementation of pelvic magnetic resonance imaging-only planning for external beam radiation therapy. *International Journal of Radiation Oncology, Biology, Physics*. 105(3): 479–492.

Bruynzeel, A., Tetar, S., Oei, S., et al. 2019. A prospective single-arm phase 2 study of stereotactic magnetic resonance guided adaptive radiation therapy for prostate cancer: Early toxicity results. *International Journal of Radiation Oncology, Biology, Physics*. 105(5): 1086–1094.

Byrne, M., Archibald-Heeren, B., Hu, Y., et al. 2022. Varian ethos online adaptive radiotherapy for prostate cancer: Early results of contouring accuracy, treatment plan quality, and treatment time. *Journal of Applied Clinical Medical Physics*. 23: e13479.

Byrne, M., Meei, A., Archibald-Heeren, B., et al. 2023. Intra-fraction motion and margin assessment for Ethos online adaptive radiotherapy treatments of the prostate and seminal vesicles. *Advances in Radiation Oncology*. 9: 101405.

Calmels, L., Sibolt, P, Astrom, L., et al. 2023. Evaluation of an automated template-based treatment planning system for radiotherapy of anal, rectal and prostate cancer. *Technical Innovations & Patient Support in Radiation Oncology*. 22: 30–36.

Chen, G., Tai, A., Puckett, L., et al. 2021. Clinical implementation and initial experience of real-time motion tracking with jaws and multileaf collimator during helical tomotherapy delivery. *Practical Radiation Oncology*. 11: e486–e495.

Christiansen, R., Dysager, L., Ronn, C., et al. 2022. Online adaptive radiotherapy potentially reduces toxicity for high-risk prostate cancer treatment. *Radiotherapy and Oncology*. 167: 165–171.

Chrystall, D., Mylonas, A., Hewson, E., et al. 2023. Deep learning enables MV-based real-time image guided radiation therapy for prostate cancer patients. *Physics in Medicine and Biology*. 68: 095016.

Colvill, E., Booth, J., Nill, S., et al. 2016. A dosimetric comparison of real-time adaptive and non-adaptive radiotherapy: A multi-institutional study encompassing robotic, gimbaled, multileaf collimator and couch tracking. *Radiotherapy and Oncology*. 119: 159–165.

Dassen, M., Janssen, T., Kusters, M., et al. 2023. Comparing adaptation strategies in MRI-guided online adaptive radiotherapy for prostate cancer: Implications for treatment margins. *Radiotherapy and Oncology*. 186: 109761.

De Cock, L. Draulans, C., Pos, F., et al. 2023. From once-weekly to semi-weekly whole prostate gland stereotactic radiotherapy with focal boosting: Primary endpoint analysis of the multicentre phase II hypo-FLAME 2.0 trial. *Radiotherapy and Oncology*. 185: 109713.

De Roover, R., Crijns, W., Poels, K., et al. 2021. Automated treatment planning of prostate stereotactic body radiotherapy with focal boosting on a fast-rotating O-ring linac: Plan quality comparison with C-arm linacs. *Journal of Applied Clinical Medical Physics*. 22: 59–72.

Draulans, C., van der Heide, U., Haustermans, K., et al. 2020. Primary endpoint analysis of the multicentre phase II hypo-FLAME trial for intermediate and high risk prostate cancer. *Radiotherapy and Oncology*. 147: 92–98.

Ferro, M., Cobelli, O., Musi, G., et al. 2022. Radiomics in prostate cancer: An up-to-date Review. *Therapeutic Advances in Urology*. 14: 1–37.

Gallio, E., Sardo, A., Badellino, S., et al. 2023. Helical tomotherapy and two types of volumetric modulated arc therapy: Dosimetric and clinical comparison for several cancer sites. *Radiological Physics and Tehcnology*. 16: 272–283.

Gaudreault, M., Chang, D. Hardcastle, N., et al. 2023. Combined biology-guided radiotherapy and Lutetium PSMA theranostics treatment in metastatic castrate-resistant prostate cancer. *Frontiers in Oncology*. 13: 1134884.

Gaudreault, M., Chang, D., Hardcastle, N., et al. 2022. Combined biology-guided radiotherapy and lutetium PSMA treatment in metastatic prostate cancer. *International Journal of Radiation Oncology, Biology, Physics*. 114(3): S116.

Ghadjar, P., Fiorino, C., af Rosenschold, P., et al. 2019. ESTRO ACROP consensus guideline on the use of image guided radiation therapy for localised prostate cancer. *Radiotherapy and Oncology*. 141: 5–13.

Gonzalez-Moya, A., Dufreneix, S. Ouyessad, N., et al. 2021. Evaluation of a commercial synthetic computed tomography generation solution for magnetic resonance imaging-only Radiotherapy. *Journal of Applied Clinical Medical Physics*. 22(6): 191–197.

Gorovets, D., Burleson, S., Jacobs, L., et al. 2020. Prostate SBRT with intrafraction motion management using a novel linear accelerator-based MV-kV imaging method. *Practical Radiation Oncology.* 10: e388–e396.

Gregoire, V., Guckenberger, M., Haustermans, K., et al. 2020. Image guidance in radiation therapy for better cure of cancer. *Molecular Oncology.* 14: 1470–1491.

Hackett, S., van Asselen, B., Philippens, M., et al. 2022. Chapter 6: Magnetic resonance-guided adaptive radiotherapy: Technical concepts. In *Image-guided high-precision radiotherapy.* Ed. E. Troost. Cham: Springer Nature, 135–158.

Hales, R., Rodgers, J. Whiteside, L., et al. 2020. Therapeutic radiographers at the helm: Moving towards radiographer-led MR-guided radiotherapy. *Journal of Medical Imaging and Radiation Sciences.* 51: 364–372.

Johnstone, E., Wyatt, J., Henry, A., et al. 2018. Systematic review of synthetic computed tomography generation methodologies for use in magnetic resonance imaging-only radiation Therapy. *International Journal of Radiation Oncology, Biology, Physics.* 100(1): 199–217.

Kashani, R., Olsen, J. 2018. Magnetic resonance imaging for target delineation and daily treatment modification. *Seminars in Radiation Oncology.* 28: 178–184.

Keall, P., Ng, J., Juneja, P., et al. 2016. Real-time 3D image guidance using a standard LINAC: Measured motion, accuracy, and precision of the first prospective clinical trial of Kilovoltage intrafraction monitoring-guided gating for prostate cancer radiation therapy. *International Journal of Radiation Oncology, Biology, Physics.* 94(5)1015–1021.

Keall, P., Nguyen, D., O'Brien, R., et al. 2018. The first clinical implementation of real-time image-guided adaptive radiotherapy using a standard linear accelerator. *Radiotherapy and Oncology.* 127: 6–11.

Keall, P., Poulsen, P., Booth, J. 2019. See, think, and act: Real-time adaptive radiotherapy. *Seminars in Radiation Oncology.* 29: 228–235.

Ketcherside, T., Shi, C., Chen, Q., et al. 2023. Evaluation of repeatability and reproducibility of radiomic features produced by the fan-beam kV-CT on a novel ring gantry-based PET/CT linear accelerator. *Medical Physics.* 50: 3719–3725.

Kirby, M., Calder, K-A. 2019. *On-treatment verification imaging: A study guide for IGRT.* Boca Raton, FL: CRC Press, Taylor & Francis Group.

Kishan, A., Ma, T., Lamb, J., et al. 2023. Magnetic resonance imaging-guided vs computed tomography-guided stereotactic body radiotherapy for prostate cancer: The MIRAGE randomised clinical trial. *Jama Oncology.* 9(3): 365–373.

Kuo, H., Della-Biancia, C., Damato, A., et al. 2023. Clinical experience and feasibility of using 2D-kVimage online intervention in the ultrafractionated stereotactic radiation treatment of prostate cancer. *Practical Radiation Oncology.* 13: e308–e318.

Ladbury, C., Amini, A., Schwer, A., et al. 2023. Clinical applications of magnetic resonance-guided radiotherapy: A narrative review. *Cancers.* 15: 2916.

Lavrova, E., Garrett, M., Wang, Y-F., et al. 2023. Adaptive radiation therapy: A review of CT-based techniques. *Radiology: Imaging Cancer.* 5(4): e230011.

Lawes, R. Barnes, H. Herbert, T., et al. 2022. MRI-guided adaptive radiotherapy for prostate cancer: When do we need to account for intra-fraction motion?. *Clinical and Translational Radiation Oncology.* 37: 85–88.

Leeman, J., Cagney, D., Mak, R., et al. 2022. Magnetic resonance–guided prostate stereotactic body radiation therapy with daily online plan adaptation: Results of a prospective phase 1 trial and supplemental cohort. *Advances in Radiation Oncology.* 7: 100934.

Liu, H., Schall, D., Curry, H., et al. 2023. Review of cone beam computed tomography based online adaptive radiotherapy: Current trend and future direction. *Radiation Oncology.* 18: 144.

Loblaw, A. 2020. Ultrahypofractionation should be a standard of care option for intermediate-risk prostate cancer. *Clinical Oncology.* 32: 170–174.

Lovelock, M., Messineo, A., Cox, B., et al. 2015. Continuous monitoring and intrafraction target position correction during treatment improves target coverage for patients undergoing SBRT prostate therapy. *International Journal of Radiation Oncology, Biology, Physics.* 91(3): 588–594.

Lu, L., Chao, E., Zhu, T., et al. 2023. Sequential monoscopic image-guided motion compensation in tomotherapy stereotactic body radiotherapy (SBRT) for prostate cancer. *Medical Physics.* 50: 518–528.

Ma, T. Ladbury, C., Tran, M., et al. 2023. Stereotactic body radiation therapy: A radiosurgery society guide to the treatment of localized prostate cancer illustrated by challenging cases. *Practical Radiation Oncology.* In Press.

Ma, T., Lamb, J., Casado, M., et al. 2021. Magnetic resonance imaging-guided stereotactic body radiotherapy for prostate cancer (mirage): A phase III randomized trial. *Bmc Cancer.* 21: 538.

Maas, J., Dobelbower, M., Yang, E., et al. 2023. Prostate stereotactic body radiation therapy with a focal simultaneous integrated boost: 5-year toxicity and biochemical recurrence results from a prospective trial. *Practical Radiation Oncology.* 13: 466–474.

Mannabe, Y., Hashimoto, S., Mukouyama, H., et al. 2021. Stereotactic body radiotherapy using a hydrogel spacer for localized prostate cancer: A dosimetric comparison between tomotherapy with the newly-developed tumor-tracking system and Cyberknife. *Journal of Applied Clinical Medical Physics.* 22: 66–72.

McComas, K., Yock, A., Darrow, K., et al. 2023. Online adaptive radiation therapy and opportunity cost. *Advances in Radiation Oncology.* 8: 101034.

McNair, H. Joyce, E., O'Gara, G., et al. 2021. Radiographer-led online image guided adaptive radiotherapy: A qualitative investigation of the therapeutic radiographer role. *Radiography.* 27: 1085–1093.

McNair, H., Wiseman, T., Joyce, E., et al. 2020. International survey; current practice in On-line adaptive radiotherapy (ART) delivered using Magnetic Resonance Image (MRI) guidance. *Technical Innovations and Patient Support in Radiation Oncology.* 16: 1–9.

McPartlin, A., Li, X., Kershaw, L., et al. 2016. MRI-guided prostate adaptive radiotherapy – A systematic review. *Radiotherapy and Oncology.* 119: 371–380.

Moazzezi, M., Rose, B., Kisling, K., et al. 2021. Prospects for daily online adaptive radiotherapy via ethos for prostate cancer patients without nodal involvement using unedited CBCT auto-segmentation. *Journal of Applied Clinical Medical Physics.* 22: 82–93.

Montalvo, S., Meng, B., Lin, M-H., et al. 2022. Case report: Adaptive radiotherapy in the radiation salvage of prostate cancer. *Frontiers in Oncology.* 12: 898822.

Morgan, H., Wang, K., Yan, Y., et al. 2023. Preliminary evaluation of PTV Margins for online adaptive radiation therapy of the prostatic fossa. *Practical Radiation Oncology.* 13: e345–e353.

Murray, J. Tree, A. 2019. Prostate cancer – Advantages and disadvantages of MR-guided RT. *Clinical and Translational Radiation Oncology.* 18: 68–73.

Oderinde, O., Han, C., Sun, Z., et al. 2022. Feasibility and dosimetric benefits of adaptive planning in prostate cancer radiotherapy using a novel treatment Planning machine with integrated dual kVCT/pet imaging systems. *International Journal of Radiation Oncology, Biology, Physics.* 114: E592.

Oderinde, O., Shirvani, S., Olcott, P., et al. 2021. The technical design and concept of a PET/CT linac for biology-guided radiotherapy. *Clinical and Translational Radiation Oncology.* 29: 1–7.

Oehler, C., Roehner, N., Sumila, M., et al. 2022. Intrafraction prostate motion management for ultra-hypofractionated radiotherapy of prostate cancer. *Current Oncology.* 29: 6314–6324.

Oldenburger, E., de Roover, R., Poels, K., et al. 2023. "Scan-(pre)Plan-treat" workflow for bone metastases using the ethos therapy system: A single-center, Insilico experience. *Advances in Radiation Oncology.*8: 101258.

Ong, W., Cheung, P., Chung, H., et al. 2023. Two-fraction stereotactic ablative radiotherapy with simultaneous boost to MIR-defined dominant intra-prostatic lesion – Results from the 2SMART phase 2 trial. *Radiotherapy and Oncology.* 181: 109503.

Pasquier, D., Nickers, P., Peiffert, D., et al. 2023. Intrafraction motion during CyberKnife® prostate SBRT: Impact of imaging frequency and patient factors. *European Urology Open Science.* 54: 80–87.

Pathmanathan, A., van As, N., Kerkmeijer, L., et al. 2018. Magnetic resonance imaging-guided adaptive radiation therapy: A "Game Changer" for prostate treatment? *International Journal of Radiation Oncology, Biology, Physics.* 100(2): 361–373.

Poon, D., Yang, B., Geng, H., et al. 2022a. Analysis of online plan adaptation for 1.5T magnetic resonance-guided stereotactic body radiotherapy (MRgSBRT) of prostate cancer. *Journal of Cancer Research and Clinical Oncology.* Published online 24 February 2022.

Price, A., Schiff, J., Laugeman, E., et al. 2023. Initial clinical experience building a dual CT- and MR-guided adaptive radiotherapy program. *Clinical And Translational Radiation Oncology.* 42: 100661.

RCR (Royal College of Radiologists). 2021. *On-target 2: Updated guidance for image-guided radiotherapy.* London: The Royal College of Radiologists.

Rose, C., Mukwada, G., Skorska, M., et al. 2023. Intrafraction motion during CyberKnife® prostate SBRT: Impact of imaging frequency and patient factors. *Springer Nature.* Preprint.

Sibolt, P., Andersson, L., Calmels, L., et al. 2021. Clinical implementation of artificial intelligence-driven cone-beam computed tomography-guided online adaptive radiotherapy in the pelvic region. *Physics and Imaging in Radiation Oncology.* 17: 1–7.

Smith, G., Dunlop, A., Alexander, S., et al. 2023. Evaluation of therapeutic radiographer contouring for magnetic resonance image guided online adaptive prostate radiotherapy. *Radiotherapy and Oncology*. 180: 109457.

Sonke, J-J., Aznar, M., Rasch, C., et al. 2019. Adaptive radiotherapy for anatomical changes. *Seminars in Radiation Oncology*. 29: 245–257.

Speight, R., Dubec, M., Eccles, C., et al. 2021. IPEM topical report: Guidance on the use of MRI for external beam radiotherapy treatment planning. *Physics In Medicine and Biology*. 66: 055025.

Sritharan, K., Tree, A. 2022. MR-guided radiotherapy for prostate cancer: State of the art and future perspectives. *British Journal of Radiology*. 95: 20210800.

Tegtmeier, R., Ferris, W., Bayouth, J., et al. 2022. Characterization of imaging performance of a novel helical kVCT for use in image-guided and adaptive radiotherapy. *Journal of Applied Clinical Medical Physics*. 23: e13648.

Tetar, S., Bruynzeel, A., Oei, S., et al. 2021. Magnetic resonance-guided stereotactic radiotherapy for localized prostate cancer: Final results on patient-reported outcomes of a prospective phase 2 study. *European Urology Oncology*. 4: 628–634.

Teunissen, F., Willigenburg, T., Tree, A., et al. 2023. Magnetic resonance-guided adaptive radiation therapy for prostate cancer: The first results from the MOMENTUM study—An international registry for the evidence-based introduction of magnetic resonance-guided adaptive radiation therapy. *Practical Radiation Oncology*. 13: e261–269.

Tocco, B., Kishan, A., Ma, T., et al. 2020. MR-guided radiotherapy for prostate cancer. *Frontiers in Oncology*. 10: 616291.

Tudor, G., Bernstein, D., Riley, S., et al. 2020. *Geometric Uncertainties in Daily Online IGRT: Refining The CTV-PTV Margin For Contemporary Photon Radiotherapy*. London: The British Institute of Radiology.

Turkkan, G., Bilici, N., Sertel, H., et al. 2022. Clinical utility of a 1.5T magnetic resonance imaging-guided linear accelerator during conventionally fractionated and hypofractionated prostate cancer radiotherapy. *Frontiers in Oncology*. 12: 909402.

Van As, N., Tree, A., Patel, J., et al. 2023. 5-year outcomes from PACE-B: An international phase III randomised controlled trial comparing stereotactic body radiotherapy (SBRT) vs. conventionally fractionated or moderately hypo fractionated external beam radiotherapy for localised prostate cancer. *International Journal of Radiation Oncology, Biology, Physics*. 117(4): e2–e3.

Ward, A., Martinou, M., Kidane, G., et al. 2022. Daily adaptive radiotherapy using the Varian ETHOS system to improve dose distribution during treatment to the upper abdomen. *Clinical Oncology*. 34: e6–e11.

Waters, M. Price, A., Laugeman, E., et al. 2024. CT-based online adaptive radiotherapy improves target coverage and organ at risk (OAR) avoidance in stereotactic body radiation therapy (SBRT) for prostate cancer. *Clinical and Translational Radiation Oncology* 44: 100693.

Westley, R., Biscombe, K., Dunlop, A., et al. 2023. Interim toxicity analysis from the randomised HERMES trial of 2- and 5- fraction magnetic resonance imaging-guided adaptive prostate radiation therapy. *International Journal of Radiation Oncology, Biology, Physics*. 118(3): 682–687.

Winkel, D., Bol, G., Kroon, P., et al. 2019. Adaptive radiotherapy: The Elekta Unity MR-linac concept. *Clinical and Translational Radiation Oncology*. 18: 54–59.

Wohlfahrt, P., Schellhammer, S. 2022. Chapter 4: In-room systems for patient positioning and motion control. In *Image-guided high-precision radiotherapy*. Ed. E. Troost. Cham: Springer Nature, 91–107.

Wyatt, J. Pearson, R., Frew, J., et al. 2023a. The first patients treated with MR-CBCT soft-tissue matching in a MR-only prostate radiotherapy pathway. *Radiography*. 29: 347–354.

Wyatt, J., Kaushik, S., Cozzini, C., et al. 2023b. Comprehensive dose evaluation of a Deep Learning based synthetic Computed Tomography algorithm for pelvic Magnetic Resonance-only radiotherapy. *Radiotherapy and Oncology*. 184: 109692.

Zou. W., Dong, L., Teo, B-K. 2018. Current state of image guidance in radiation oncology: Implications for PTV margin expansion and adaptive therapy. *Seminars in Radiation Oncology*. 28: 238–247.

Zwart, L., Ong, F., ten Asbroek, L., et al. 2022. Cone-beam computed tomography-guided online adaptive radiotherapy is feasible for prostate cancer patients. *Physics and Imaging in Radiation Oncology*. 22: 98–103.

6 Particle Therapy

6.1 INTRODUCTION

Aspects of proton beam therapy and image guidance during it are introduced in Kirby and Calder (2019, section 10.6). Within that section, proton beam therapy is introduced (section 10.6.1); established and developing particle therapy platforms are discussed (section 10.6.2); clinical utility and some of the clinical challenges associated with proton beam therapy are presented (some of which are also pertinent for the treatment and image guidance for prostate cancer (PCa) patients) (sections 10.6.3 and 10.6.4); and image guidance equipment (X-ray-based and nonionising-radiation-based) for planar and volumetric on-treatment imaging, using gantry-mounted, in-room and independent systems, is discussed (section 10.6.5). The reader is encouraged to re-acquaint themselves with these points in complement to the work in this chapter.

6.2 THE CASE FOR PROTON BEAM THERAPY FOR PROSTATE CANCER PATIENTS

Recent conferences and publications have seen a rise in support for the use of proton beam therapy (PBT) for PCa. Notable journals like the Red Journal (https://www.redjournal.org/) and, more recently, the British Journal of Radiology (https://www.birpublications.org/toc/bjr/current) have run special issues on particle therapy and also innovations in PCa, highlighting the current and future developments in the use of PBT (Wu and Fan 2022). Among the evidence accruing, including (i) five-year cancer progression-free survival rates reaching 99% in low- and intermediate-risk patients; (ii) five-year cancer progression-free survival rates of over 70% in high-risk patients; (iii) minimal bowel and urinary side effects, returning to pre-treatment or improved pre-treatment levels within two years of treatment; (iv) minimal impact on sexual function, with a decrease in erectile function in the first year, stabilising in years 2 and 3; and (v) mirroring of patient- and provider-reported data concurring with study results across institutions.

Like other external beam radiotherapy (EBRT) techniques, PBT is used to try to increase the therapeutic index and minimise normal tissue irradiation – with a natural dosimetric advantage over photon techniques of having little/no exit dose within the beam and therefore exploiting some of the physical properties of the proton beam to reduce normal tissue and integral dose (see Section 6.6 below) below those of standard conformal photon techniques like intesnity modulated radiotherapy (IMRT) or volumetric modulated arc therapy (VMAT).

The use of PBT for treating prostate cancer is controversial, though, largely because of the higher treatment costs, the success of modern photon techniques and other approaches, and also the uncertainty of any clinically significant benefit. The distinct benefits in physical dose distributions have been widely demonstrated for PBT as has the development of intensity-modulated proton therapy (IMPT), with clear reductions in urinary bladder and rectal doses in dosimetric modelling studies (Wu and Fan 2022). There are questions, though, in terms of doses to the femoral heads – especially for the use of single or parallel opposed lateral beams, with some studies (e.g., Trofimov et al. 2007; Georg et al. 2014; Scobioala et al. 2016) failing to demonstrate clear advantages of PBT, and some suggestions that it may even be disadvantageous. However, more optimised beam angles help to reduce femoral head dose (Tang et al. 2012), as does the use of novel gantry techniques like arc PBT (Wu and Fan 2022).

Because of the relatively long survival rates and excellent disease outcomes for the majority of PCa patients, PBT could be considered as a natural treatment option, especially with the reduction

DOI: 10.1201/9781003050988-6

in risk of secondary cancer induction (see Section 6.6) because of the lower integral dose, a risk which could be reduced, compared with IMRT, by nearly 40% in some simulated cases (Wu and Fan 2022).

In their recently published review of PBT use in newly diagnosed cancer patients (in general) in the US, Nogueira et al. (2022) categorised non-metastatic prostate cancer as a Group 2 indication for PBT, in that there is still a need for continued development in clinical evidence and comparative effectiveness for its appropriate use, especially when one considers current photon techniques and the clinical outcomes coming from, in particular, ultrahypofractionated techniques and the onset of online adaptive radiotherapy (ART) (see Chapters 5, 8 and 9). The classification is defined as Coverage for Evidence Development (CED) and is evolving particularly for prostate patients, where further data are needed to understand PBT effectiveness compared to IMRT/VMAT and brachytherapy. PCa is the most common site for Group 2 indications, and primary treatment with PBT is only recommended in the context of a prospective clinical trial or registry, as PBT may not be of added benefit in terms of GI toxicity for PCa (Nogueira et al. 2022; ASTRO 2017). In the UK, PCa is not supported in the initial NHS England commissioning of the national high-energy PBT service, developed in Manchester and London (Burnet et al. 2020), as there is little scientific, clinical or economic value seen in running studies which will probably show equivalence in terms of tumour control or toxicity.

In Hoffman's (2021) article presented in that year's ASTRO meeting, the evolution of PBT over time was described, with pencil beam scanning techniques making more conformal dose distributions possible (especially proximal to the target volume in the proton beam direction) and also enabling IMPT, with greater rectum sparing and potential further benefits from the lower integral dose with theoretically lower expectations of radiation-induced secondary cancers. PBT has been shown to deliver lower doses to the testicles, with maintenance of testosterone levels after therapy (compared to some decreases in levels for IMRT). But the clinical relevance and significance are still to be proven

Patient- and physician-reported outcomes have shown equivalent or lower rates of early and late gastrointestinal (GI) and genito-urinary (GU) toxicity when compared with conventional fractionation photon-based EBRT, but, in many cases, the patient numbers are naturally relatively small, especially when compared with photon treatments. Hypofractionation brings the potential for different results (for both photons and protons), and also a potential answer for PBT with lower costs. Pooled Phase II studies of hypofractionated PBT suggest higher rates of late Grade 2+ GI toxicity with moderate hypofractionation (compared with photon therapy), but no differences in acute Grade 2+ GI or acute/late Grade 2+ GU toxicity. Larger case series, from studies like those conducted at the University of Florida, indicate promising tumour control with reasonable biochemical failure rates (Hoffman 2021).

Researchers like Ojerholm and Bekelman (2018) and Poon et al. (2022) push forward the arguments for PBT for PCa patients – the challenges that are faced, but the opportunities too. PBT is safe and effective for prostate treatment – although some question whether it adds value (Wu and Fan 2022). By value, one could consider the general improvement of health outcomes (either better tumour control or fewer side effects). But PBT is unlikely to improve control: few PCa patients die from lower-risk disease, but dose escalation beyond current standards may still be unsafe (although ultra-hypofractionated (UHF) studies show promise here; see Chapter 8). Value may be principally focussed on meaningful reduction of GU and GI side effects and erectile dysfunction, although results are mixed for high-dose studies (Hoffman 2021; Wu and Fan 2022; Poon et al. 2022). Value also comes from the reduction in total body exposure, with the potential for less risk of secondary cancer induction. Modelling studies indicate this could be a 25–50% reduction in risk for PBT vs IMRT, as a result of a consistently smaller irradiation volume of normal tissues. But secondary cancers are rare for prostate treatments, so it is unclear whether this would be a distinct clinical advantage (Ojerholm and Bekelman 2018; Poon et al. 2022).

6.3 THE NEED FOR IMAGE GUIDANCE.

Whereas the rapid fall-off in dose distally beyond the Bragg peak within the proton beam is a distinct advantage over photon beams, it also warrants a more stringent image guidance for accurate placement of the high-dose volume within the patient. Although a lack of divergence of the beam makes set-up errors in the beam direction less of an issue, proton beams are more sensitive to changes in the anatomy in the beam direction, especially proximal to the spread-out Bragg peak, than photon techniques (Wu and Fan 2022; Kirby and Calder 2019). These are often described as changes in the Water Equivalent Thickness (WET) in the beam path; changes here compared with those at the time of pre-treatment scanning/planning can lead to inconsistent dose coverage of the planning target volume (PTV) around the prostate from the proton beam(s) (Hoffman 2021; Kirby and Calder 2019; ASTRO 2017).

For prostate treatments in particular, variations in WET can be caused by patient surface changes, displacement of dense tissues in the proximal part of the beam (e.g., positioning of the femoral heads and bony structures for certain beam directions) or variation in low-density regions like the rectum, bowel and bladder – again, dependent upon beam direction in planning and also consistency in bowel/bladder preparation (Wu and Fan 2022).

Most centres have relied upon 2D planar kV imaging for image guidance in PBT facilities in combination with fiducial markers (FMs), although some study results are still reported from the use of bony anatomy for on-treatment verification imaging (Takagi et al. 2021). The use of bony landmarks is not recommended for modern photon techniques and, even with the use of FMs, there is limited information to help assess changes in WET, which could lead to range uncertainty (see Section 6.5).

Ideal image guidance comes from volumetric techniques, especially kV cone-beam computed tomography (CBCT), prominent in photon therapy, but still a step behind in some PBT facilities. In-room computed tomography (CT) is used for volumetric imaging in some PBT facilities. Volumetric techniques facilitate daily assessment of WET changes, dose reconstruction, adaptive online processes and increases in the robustness of PBT dose delivery and subsequent dose distributions, as well as the usual online adjustments for set-up errors. Volumetric methods are necessary for accurate beam and dose placement; especially for PBT and further advanced techniques like IMPT. The move towards magnetic resonance (MR) guidance for on-treatment image guidance is being developed for PBT facilities and should further aid the techniques used for prostate treatments (Kirby and Calder 2019).

The benefits may still not be fully realised – due to improper dose delivery as a result of uncertainties compromising the application of dose conformity; the geometric positioning of high dose gradients in dose distributions which are possible and beneficial regarding surrounding normal tissues, which improved image guidance would help with (Poon et al. 2022; Kirby and Calder 2019). Major contributors to the current limited potential of PBT could be discrepancies, including uncertainties in treatment planning and delivery. Interfraction motion (analysed using CT) showed higher uncertainties for interfraction range changes in lateral beams compared with anterior, likely due to the relative positions of the femoral head and prostate between fractions for PBT (Poon et al. 2022). This is something that is not always possible to fully detect with 2D planar imaging with orthogonal images for on-treatment image guidance. A consequence of the range uncertainty is translation into relatively large margins – 3.5% of the range plus an additional 2–3 mm (compare those with those discussed in Chapter 7), compromising the dosimetric benefits of PBT, with the uncertainties also occurring due to tissue density changes in the beam path (WET) due to patient set-up or anatomy differences occurring inter- and intrafractionally (Poon et al. 2022; Hoffman, 2021; Kirby and Calder 2019; ASTRO 2017).

Albertini and colleagues (2020) have discussed the use of online ART for PBT because a single plan generated some time before treatment starts is likely to be insufficient to represent the daily dose to the target and organs at risk if there are significant changes in patient positioning, shape, and internal anatomical positions before and during the treatment course. This is particularly true and an issue for PBT due to the finite range of the beams. From the advantage of the steep dose gradients possible with PBT (and needed for the best dose conformity and therapeutic index), comes an essential need to tighten the dependence of the delivered dose on the range accuracy (Albertini et al. 2020).

One of the most significant sources of range inaccuracy is an outdated anatomical profile. Anatomical changes in the patient alter the position of the Bragg peak (and therefore the position of the high-dose volume), changing the dose distribution, potentially leading to dose nonhomogeneity in the target and/or overdosage of organs at risk (OARs)/normal tissues (Mackay 2018). For this reason, plan adaptation is required as soon as anatomical variations occur (best detected using volumetric imaging methods), ideally online.

Daily volumetric imaging, development of fast online planning and quality assurance (QA) techniques and *in-vivo* range measurements (e.g., positron emission tomography (PET) and prompt gamma imaging) and verification are image guidance methods which would serve this purpose. In-room CT methods are those presently in use; as well as on-board kV orthogonal and volumetric CBCT equipment, which are becoming standards of use (Mackay 2018). Additional modes of imaging associated with directly assessing the position of dose deposition – termed range verification – are being developed (see Section 6.5 below). In the future imaging and computed tomography with the proton beam itself will become available and MR guidance is likely to become both possible and a standard for plan adaptation for PBT for prostate treatments (Albertini et al. 2020; Kirby and Calder 2019; Mackay 2018).

6.4 PRE-TREATMENT PROCEDURES FOR PROSTATE CANCER RADIOTHERAPY

The ASTRO (2017) policy guidance nicely summarises the pre-treatment procedures needed for planning PCa treatments with PBT, many of which are similar to those for photon treatments (see Kirby and Calder 2019, chapters 7 and 8). Simulation and imaging are done through 3D image acquisition with CT, PET/CT and/or MR imaging being essential prerequisites. Bladder and bowel preparation are needed for consistency of patient positioning interfractionally; also, consistent and appropriate immobilisation for hips/femurs are needed to prevent possible interfractional anatomical changes (especially those that could affect the WET in the proton beam paths chosen during planning) (see Chapter 2). The impact of immobilisation on high-dose volume placement (because of the nature of particle beams) must be considered for the immobilisation system used and for the selection of beam directions; photon-based immobilisation equipment is not always appropriate for use in PBT (ASTRO 2017; Wu and Fan 2022; Hoffman 2021).

Careful consideration must be given to the contouring of appropriate volumes like the gross tumour volume (GTV), clinical target volume (CTV) and the generation of the planning target volume (PTV). Particular to PBT, target expansion in the beam direction must also cover uncertainties in proton beam range, which may make the generation of a single PTV difficult. Coverage of the CTV is essential in PBT planning, especially in the presence of likely uncertainties and possible changes in WET. Normal tissues and OARs are contoured as per usual (Wu and Fan 2022; Poon et al. 2022; Hoffman 2021).

In radiation dose prescribing, typical prescription may define a dose of at least 99% of the target dose to the CTV. Dose constraints to OARs should be evaluated in the presence of delivery and range uncertainties for PBT. With dosimetric planning and calculations, while being likely more conformal, PBT plans are more susceptible to uncertainties in patient positioning and/or proton range *in vivo*. Nominal treatment plans should have a robust approach, examining the possibilities of likely positional and range uncertainties to ensure that the planned CTV is suitably covered in each fraction and that normal tissue sparing is also maintained in the presence of expected/possible range uncertainties for prostate treatments (which are dependent upon the chosen beam directions) (ASTRO 2017; Wu and Fan 2022; Hoffman 2021; Poon et al. 2022; Kirby and Calder 2019).

Tilbaek et al. (2023), in their recent RCT, highlighted the need for a comfortably filled bladder and minimal air cavities in the rectum (with the use of a catheter/laxatives, if excessive); if still excessive, CT numbers could be overridden in the planning CT scan with water or surrounding soft tissue CT number. This could be considered a robust method, since there is little likelihood of bowel gas being precisely replicated interfractionally. Planning is assessed on both the original and overridden CT numbers to assess the robustness of the plan.

Plans used were four field plans – two post obliques and two lateral obliques – in order to avoid beams going directly through the femoral heads. Spot scanning algorithms and robust multifield optimisation were also used (Tilbaek et al. 2023).

Forsthoefel et al. (2022) discussed their methods of using pre-treatment CT and MR scanning, a comfortably full bladder, and empty rectum. CT-MR fusion was used for contouring; opposed lateral beams were used for planning and a robust evaluation was performed for six directional positional shifts of 5 mm and radiation uncertainty of 3.5% to ensure coverage of CTV. For delivery, one lateral field was delivered per day to assist in patient and centre day-to-day logistics. Single field optimisation was used for all PBT plans.

Poon et al. 2022 discussed a simple bilateral field arrangement which is most used for PBT for localised prostate cancer. Other arrangements (e.g., oblique posterior and oblique laterals (Tilbaek et al. 2023)) can be used to spare doses to the femoral heads or even anterior-posterior or anterior oblique fields for patients with hip prostheses or previously irradiated hips. Pencil beam scanning (PBS) systems demonstrate a tighter penumbra than passive scattering delivery systems, and IMPT brings even further improvements in dose conformity, with the potential for better protection of critical organs (e.g., bladder and rectum for prostate cancer), thereby potentially reducing toxicity (Poon et al. 2022).

Dosimetric advantages can be further enhanced with the use of rectal spacers and robust planning optimisation, used to account for uncertainties which cannot be corrected (e.g., movement of target and OARs during treatment), and are in routine use for planning PBT treatments for the prostate. These techniques help predict and illustrate the actual dose likely delivered to the patient, which could further improve dose optimisation and clinical decision making (Poon et al. 2022).

MR-only workflows (pre-treatment) and MR on-treatment guidance are being developed (Kirby and Calder 2019, chapter 10; Hoffman 2021), with simulations being used to investigate (1) the effect of PBS on the MR magnetic field homogeneity for scanning and (2) the deflection of the proton beam trajectories in reaching the treatment volume. Work is developing to limit both these effects. With regard to pre-treatment planning, dose distribution changes have been found to be generally small for magnetic field strengths of up to 0.5 T, and generally correctable (using the PBS equipment and other software-based correction methods) up to 1.5 T (Hoffman 2021; Kirby and Calder 2019, chapter 10).

MR-only workflow (pre-treatment) requires methods for dose calculations, for which there are four main categories of MRI-to-CT conversion techniques: (a) bulk density, (b) atlas-based, (c) voxel-based and (d) deep learning. Limited work has been done so far for MR-only planning for PBT, as the difficulty lies in separating and predicting the correct bone and air hounsfield unit (HU) values on the generated pseudo or synthetic CT images. No current approach is of great maturity and completely robust, but they show promising results with clinically acceptable deviations from reference CT images and derived digitally reconstructed radiographs (DRRs). Much of the research focus is shifting to artifical intelligence (AI) and deep learning methods (see Chapter 10), which could possibly handle arbitrary inputted MR sequences and provide accurate synthetic CTs with continuous Hounsfield Unit (HU) values in very short timescales (seconds), making them ideal for online adaptive MR-guided PBT workflows (Hoffman 2021).

In Schreuder and Shamblin's (2020) horizon scanning for the future of PBT delivery, one improvement is likely to be that of arc therapy. In the same way that VMAT is now becoming a standard of care for many photon-based treatments, spot scanning arc-based PBT can be shown to be feasible and dosimetric studies demonstrate advantages over IMPT (particularly for prostate treatments) of further reduced integral dose, reduced femoral head dose and reduced dose to the OARs (rectum and bladder), with the benefit of significant reduction in beam delivery time. Unlike intensity-modulated photon treatments, IMPT and arc-based PBT beams can be modulated in three dimensions, although this would then necessitate an even greater set-up accuracy and therefore would be very highly dependent upon strict on-treatment image guidance, and an understanding/elimination of range and high-dose volume delivery uncertainties.

6.5 IMAGE-GUIDANCE EQUIPMENT AND TECHNIQUES

6.5.1 2D/Planar Imaging

The earliest developments for image guidance for PBT were from the use of 2D kV orthogonal imaging, for both fixed beam and gantry-based treatment techniques. Images are compared with the DRRs generated from the treatment plan, as a reference, with beam angles chosen at treatment planning, often corresponding with beam directions where possible (Mackay 2018; Hoffman 2021; Wu and Fan 2022; Kirby and Calder 2019). Visualising bony anatomy from the beam directions is useful to note, particularly for prostate patients, because of the effect on the proton beam range *in vivo*.

Many facilities still rely on 2D orthogonal kV energy X-ray imaging combined with implanted fiducial markers (FMs) as the soft tissue surrogate (see Section 6.5.3), although some recent studies still report the use of bony landmarks (Takagi et al. 2021). As in their use for photon treatments for prostate, 2D orthogonal imaging is limited in its use for identifying clear changes in anatomy in 3D, especially those that might lead to changes in WET and therefore to greater uncertainty in the dose distribution (range uncertainty) during PBT treatment delivery. Like photon-based 2D/planar on-treatment verification imaging, the reference images (for the planned treatment position and delivery of the planned dose distribution) are DRRs derived from the planning CT slices, and/or FM positions from the DRRs for each orthogonal view (Mackay 2018; Hoffman 2021; Wu and Fan 2022; Kirby and Calder 2019).

In some recent trial reports (some with IMPT and moderately hypofractionated regimes for high-risk patients), use is still made of daily matching of intraprostatic markers, using orthogonal 2D planar kV imaging (Choo et al. 2021, 2023; Wong et al. 2022). For volumetric data, weekly CT verification scans (over the five-week treatment course) were used to check CTV coverage with contours propagated onto the weekly CT scans to assess appropriate coverage. CBCT was not available for use on-treatment. The FMs are used for assessing prostate position and pelvic bones for assessing field placement with respect to the pelvic nodes. If coverage was deemed inadequate, a new IMPT plan should be generated.

6.5.2 3D/Volumetric Imaging

Volumetric methods (like gantry-mounted kV CBCT or in-room CT) are ideal methods for both online on-treatment image guidance for daily set-up errors and also monitoring anatomical changes interfractionally during treatment (Wu and Fan 2022; Hoffman 2021), especially the relationships between organs involved and information on interfractional anatomic changes (Poon et al. 2022; Mackay 2018; Hoffman 2021; Kirby and Calder 2019). Like photon-based techniques (for which volumetric imaging is now standard), the reference dataset for the comparison is the planning CT slices. These methods are still developing for PBT – although all newly purchased PBT facilities have gantry-mounted orthogonal kV X-ray tubes and flat panel imaging devices (AMFPIs) as standard, providing 2D/planar kV and 3D/volumetric kV CBCT options (see Figures 6.1 and 6.2), or they can have in-room CT systems ("CT-on-rails") installed.

Volumetric methods can also enable daily dose reconstruction and help facilitate daily online adaptation techniques, thereby increasing the robustness of delivery of the highly conformal PBT dose distribution with greater confidence. The in-room CT (CT-on-rails) systems provide good,

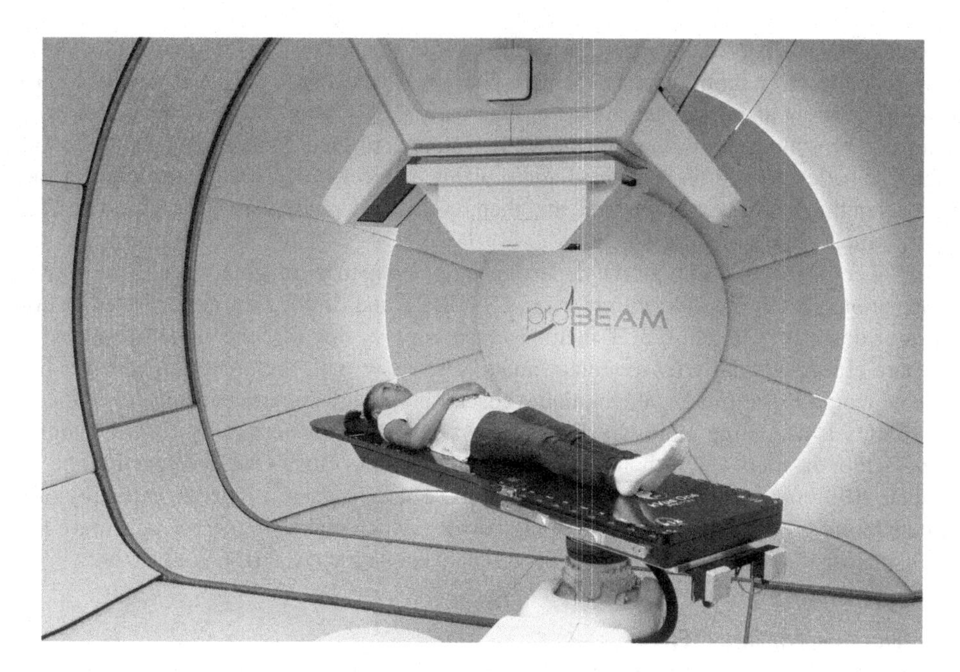

FIGURE 6.1 The Varian ProBeam PBT system with pencil beam scanning and 2D/3D kV X-ray based on-treatment IGRT. Image courtesy of Varian Medical Systems.

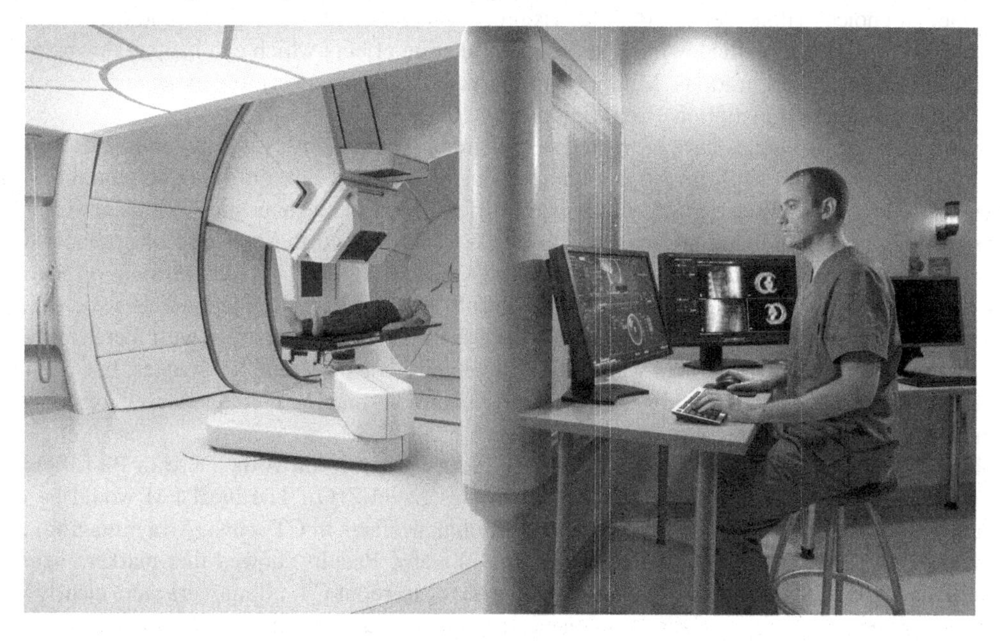

FIGURE 6.2 The Varian ProBeam PBT system with on-treatment imaging panels deployed, for 2D/planar and 3D/volumetric on-treatment IGRT. The operator is in position behind an (in-room) radiation protective screen when acquiring images and performing set-up corrections. Image courtesy of Varian Medical Systems.

high-quality (better image quality than CBCT) volumetric images (Poon et al. 2022; Albertini et al. 2020; Mackay 2018; Kirby and Calder 2019), but difficulties can be encountered in limiting patient movement between scanning and treatment (Mackay 2018; Kirby and Calder 2019). Issues of low contrast (with the system being X-ray based) and the contribution to concomitant dose are similar to those for photon-based treatments; the advent of MR is in development for volumetric image guidance for PBT (Poon et al. 2022; Hoffman, 2021; Kirby and Calder 2019).

Tilbaek et al. (2023) used daily CBCT matched to the planning CT (pCT) scan with a 6D match on bony anatomy. Dose monitoring in the pilot phase was based on CBCT and weekly control CT scans (cCTs). Patient anatomy evaluation used the daily CBCT scans, looking for large systematic deviations in prostate, rectum diameter, bladder volume, surface contour or femur position – if detected, the patient would be rescanned and replanned. Weekly cCTs were acquired for the first ten patients in the pilot phase of the trial, and then, in the RCT phase, two cCTs would be acquired during the whole course of treatment.

Albertini et al. (2020) highlight the dependence on volumetric imaging when there is a view to plan adaptation. This is possible with in-room CT, CBCT and (in the future) integrated or in-room MR. Lower doses for CT and CBCT methods ensure a lowering of concomitant dose and therefore a more efficacious daily on-treatment imaging strategy with the attraction of lower integral doses than photon-based treatment techniques (because of the dosimetric advantages of the proton beams). CBCT is becoming more widely available but still suffers from lower image quality than that provided by in-room CT, although the developments in iterative CBCT reconstruction (Kirby and Calder 2019) are providing great improvements. There is also an inherent difficulty of accurately converting greyscale to relative stopping powers, which makes it difficult to achieve full use of CBCT images for proton dose calculation and plan adaptation (Albertini et al. 2020).

6.5.3 Fiducial Markers (FMs)

As mentioned earlier, like for photon-based treatments, FMs are a popular option as soft tissue surrogates and for use with both 2D and 3D on-treatment image guidance methods. Gold FMs are the popular choice (Tilbaek et al. 2023) for visibility and contrast for CT, CBCT and MR imaging, as well as the absence of shadowing and dose-degradation effects which can happen downstream within PBT beam configurations. Migration of markers is still a concern for some, and the consequent introduction of registration errors (Poon et al. 2022).

Some of the above issues have prompted studies into marker visibility and suitability, especially for PBT. Osman et al. (2019) investigated markers using different types of imaging modalities for prostate treatments, examining gold markers (of different sizes and shapes) as well as newer polymer markers. A pelvic phantom was used to assess visibility and the presence of artefacts with imaging techniques and modalities of MR imaging, CT, CBCT, kV 2D planar, in-room orthogonal kV imaging and MV imaging (2.5 and 6 MV). Results showed that all gold markers were visible, but most produced artefacts on CT and CBCT with the magnitude increasing with increasing size of gold markers. Polymer markers were least visible and poor on lateral kV images, but polymer markers were felt to be superior to gold for the most common imaging modalities used for prostate, namely CT, CBCT and MR (Osman et al. 2019).

Reidel et al. (2022) also investigated FMs, but with a view towards their use in PBT. Several marker types were tested: gold, platinum and carbon-coated ZrO_2. The ideal FM would be one visible on daily imaging, while also producing minimal artefacts in CT scans (X-ray imaging) and inducing only slight dose perturbation for particle therapy. Results showed that markers heavier than 6 mg exhibited better contrast for X-ray images; for in-room CT, all markers were clearly visible. For fluence measurements, carbon-coated ZrO_2 markers and low-mass gold/platinum markers (0.35 mm diameter) induced perturbations 2–3 times lower than heavier gold or platinum markers of 0.5 mm diameter.

Bertholet et al. (2019), in their in-depth review of methods for real-time intrafractional motion detection, also noted the specific particle therapy-related challenges for FMs, which might require precautions. The main concern is that the high-Z material often used (gold, platinum) would cause artefacts in conventional X-ray imaging (especially CT) used for planning and image guidance. However, for PBT, inaccurate representation of electron density and HUs near inserted FMs may also result in improper dose calculations. Metal FMs can interact with particle beams (particularly scanned beams) and have an impact on therapy, influencing dose distribution, fluence and range of ions, depending on the material, thickness and location in the treatment beam. Thin markers (< 0.5 mm) or those made from relatively low-Z materials (e.g., carbon-coated zirconium DIoxide) may be considered for use in PBT in general and for the prostate in particular.

In the recent studies mentioned in Section 6.5.1 above (Choo et al. 2021, 2023; Wong et al. 2022), four carbon FMs were inserted *via* transperineal or transrectal methods 2–3 days prior to CT simulation. They were used for daily image matching with orthogonal 2D X-ray images, alongside weekly CT verification scans (over the five-week treatment course) to check CTV coverage, since CBCT was not available for use on-treatment.

6.5.4 Real-time Motion Monitoring

Mention has already been made of the in-depth review by Bertholet et al. (2019) into these methods, and they are also discussed in Chapter 5. Particularly for prostate patients, intrafractional motion is generally slower and non-periodic. Implanted FMs are common practice for some of the methods outlined in Chapter 5 for monitoring and tracking prostate motion during treatment, but the challenges for PBT as mentioned above remain for certain types of FM. Real-time imaging during particle therapy is by way of the newer installations having kV 2D planar systems; orthogonal stereoscopic kV planar imaging systems are used as well as the beam's eye view (BEV) kV imaging in some facilities (Bertholet et al. 2019)

6.5.5 MR-guided Imaging

MR guidance is discussed in detail in Kirby and Calder (2019, chapter 10), and its use for adaptive radiotherapy (ART) earlier in this book in Chapter 5. Although most PBT facilities now come equipped with volumetric kV X-ray imaging equipment and techniques, similar to those used for photon treatments, they are still limited by the inherent image contrast within the images, especially for soft-tissue definition (Kirby and Calder 2019). These inherent difficulties for X-ray imaging have been well known for MV planar, MV fan-beam and MV CBCT techniques, as well as for kV planar, kV fan-beam and kV CBCT methods (Poon et al. 2022; Morrow et al. 2012; Kirby and Calder 2019, chapter 7), as has the improvement in more easily identifying soft-tissue boundaries around the prostate compared to X-ray-based imaging (see Chapter 5).

The development of MR-guided equipment for on-treatment geometric verification and adaptive techniques is one method to counter these problems, and now it is developing well for photon LINACs (Kirby and Calder 2019, chapter 10; Chapter 5 of this book), bringing with it the possibilities for intrafractional real-time imaging as well as interfractional image guidance and adaptation. These benefits are now being developed for PBT, to harness the excellent soft-tissue-imaging possible with MR imaging with real-time information and online adaptation for the naturally more conformal dose distributions possible with PBT and the dynamic advantages of spot-scanning PBT systems. Although these are still in their infancy, and they are not without quite difficult challenges (because of the motion of charged particles (protons) in the presence of the magnetic field), the potential is there for improving the quality of PBT delivery for prostate cancer patients (Hoffman et al. 2020; Poon et al. 2022; Kirby and Calder 2019, chapter 10).

Online MRI would have the ability to offer fast real-time imaging with improved soft-tissue contrast (relative to X-ray based techniques), with the added bonus of no dose of ionising radiation, thereby lowering concomitant dose exposure still further (Poon et al. 2022; Albertini et al. 2020). In-room MRI studies are being investigated, using a robotic couch to manoeuvre the patient from imaging within the MR unit to treatment within the gantry of the PBT facility. Patient positioning accuracy of the order of 1 mm has been demonstrated to be feasible (Poon et al. 2022).

MR guidance on-treatment is being investigated and gaining momentum for both in-room and integrated systems for PBT. Some optimisation of plans and the use of PBS IMPT-style techniques can help to compensate for the effects of the inherent magnetic fields within an MR unit (especially for integrated image guidance for PBT) on charged particle trajectories. In-room MR systems present slightly less of a challenge in these respects. But the possibilities for better image guidance and also for online ART using PBT are the subject of much research (Hoffmann et al. 2020; Albertini et al. 2020; Poon et al. 2022; Kirby and Calder 2019, chapter 10).

6.5.6 Range Verification

The elements of verifying the true proton beam range *in vivo* have been mentioned earlier and are covered particularly in Parodi (2020), Poon et al. (2022), Mackay (2018) and Kirby and Calder (2019, chapter 10). QA tools are needed for assessing proton beam range *in vivo*; these include a variety of methods, including proton radiography and tomography, positron emission tomography and prompt gamma imaging techniques. On-line MR imaging may also have some verification options available (Parodi 2020; Poon et al. 2022).

Radiographic proton transmission imaging with volumetric X-ray-based CT or direct volumetric proton CT are both techniques which can be complemented by *in-vivo* PBT range verification, exploiting the detection of physical irradiation-induced secondary emissions emerging from the patient during or immediately after treatment. In particular, there are methods under development which exploit energetic gamma radiation resulting from nuclear reaction radioactive products (yielding positron emissions with subsequent annihilation, as observed in positron emission tomography (PET) imaging) and fast de-excitation radiative processes (e.g., prompt gamma (PG) emission). PG imaging for *in-vivo* range verification of PBT is still under development, but active and very promising in terms of research, where first clinical experience is being generated with full-scale collimated camera set-ups and more detector designs which can exploit different PG signatures (Parodi 2020; Poon et al. 2022; Kirby and Calder 2019, chapter 10).

Martins et al. (2021) performed a study investigating real-time PG spectroscopic range monitoring for prostate patients. The potential is to reduce further GI toxicities which are linked to a higher dose to the rectum than is planned due to anatomical variations and set-up changes (e.g., changes in WET). The new approach here was to examine the feasibility of using an endorectal balloon (ERB). ERBs have been used to stabilise the prostate during treatment but, instead of being water-filled, Martins et al. (2021) used water solutions with higher concentrations of high atomic number materials (e.g., silicon) to enable the use of the PG emissions as a dose monitor for the rectum. The PG monitoring can indicate the instances of protons striking the silicon and therefore delivering a higher dose to the rectal wall and rectum than planned.

6.5.7 Proton CT

Here we are dealing with imaging techniques that attempt to use the proton beam itself for both CT and radiographic purposes (Mackay 2018; Parodi 2020). They rely on increasing the proton beam energy beyond that usually used for treatment, so that it passes through the patient (and the peak is beyond the patient). Proton radiography uses a transmission image formed by measuring the proton position at the entrance and exit of the patient and its residual energy after transmission through the patient. Proton CT takes this further, using similar techniques employed to acquire images from different angles on a standard CT scanner, but using proton transmission projections. The resultant volumetric image is reconstructed from the measured energy loss of the protons passing through the patient.

A key limitation with the technique, especially for prostate treatments, is the proton beam energy. Most commercial PBT facilities have a maximum energy of about 250 MeV, suitable for deep-seated lesions (such as the prostate) but not high enough to pass completely through the patient with a measurable residual energy.

Research and development is underway to examine using the same radiation quality for treatment imaging, which brings potentially lower dose daily image guidance on-treatment with potentially better SPR (stopping power ratio) accuracy either from a combination of radiographic proton transmission imaging with volumetric X-ray-based CT or directly from volumetric proton CT. Low-dose proton CT shows potential for image-guided ART workflows (Parodi 2020).

6.6 SECONDARY CANCER INDUCTION

The development of secondary cancers, due to radiation treatment of the primary, is always a concern, which is why there is such emphasis on trying to lower the overall integral dose (in fact, all non-therapeutic doses, including image guidance) and concomitant doses as much as practicable (Wu and Fan 2022; Kirby and Calder 2019, chapter 5). It is one of the most serious longer-term consequences of treatments, so techniques and modalities that lower this dose and therefore the risk of secondary cancer induction, while still maintaining overall efficacy in terms of tumour control, are something to pursue. To that end, PBT would seem an obvious choice for changing techniques/modalities, especially when the life expectancy following successful prostate cancer treatment is increasing. There is a lower integral dose, dosimetrically, for PBT when compared with modern delivery techniques like VMAT for prostate, with PBT consistently demonstrating a smaller irradiated volume of normal tissue and lower integral non-target doses (Poon et al. 2022).

The risk of secondary cancer induction from photon-based techniques (EBRT and/or brachytherapy) is in the range of 0.3–0.5%, increasing to nearly 1.5% for longer-term (>10 years) follow-up. Some population studies show a 1.42-fold and 1.70-fold increase in bladder and rectal cancer development, respectively, compared with standard population rates (Wu and Fan 2022). PBT dosimetry studies naturally indicate a reduction in the overall integral dose to normal tissues as well as to the OARs. Some matched PBT/photon-based treatment studies, comparing secondary cancer incidence, have shown a lower risk in 33% of prostate cancer patients. Wider studies, including other cancer types, found PBT to be associated with a significantly lower risk than IMRT for prostate patients. Although more data are needed, the consistency with dosimetric studies seems promising (Wu and Fan 2022). From organ equivalent dose calculations, the risk of malignancies in out-of-field organs can be five times higher for IMRT than PBT for prostate treatments. Smaller irradiated volumes and lower integral doses associated with PBT may reduce risk further (Poon et al. 2022).

6.7 PBT OUTCOMES/TRIALS

There is a distinct lack of mature direct randomised controlled clinical trials making direct comparisons for efficacy of tumour control between PBT and photon-based treatment techniques in prostate cancer (Wu and Fan 2022). Studies conducted so far (but with quite dated techniques) have failed to improve tumour control or overall survival. Prospective and retrospective cohort studies have shown comparable control rate results for PBT and photon treatments, as well as for case-matched comparative studies for PBT and brachytherapy. Current evidence would suggest equivalent control rates, with no Level III evidence indicating a superiority of PBT for tumour control (Wu and Fan 2022).

Similarly, in terms of toxicity profiles, there are few prospective randomised comparative trials, but several studies have comparative data for PBT and photon treatments for prostate patients. These have included population-based studies and prospective patient-reported outcome studies (Wu and Fan 2022; Hoffman 2021). Although some studies have relatively small (<100) sample sizes, there is generally equivalent or lower and late GI and GU toxicity compared with photon-based treatments. But caution must be exercised in that sometimes endpoints differ (e.g., definition of time-points for the evaluation of acute and late effects). Some studies showed a slightly higher incidence of late GI toxicity. Although the data show a mixed landscape regarding PBT, there is general support for the potential of early GI and GU toxicity exhibiting a better profile compared with photon-based studies. For sexual function, the data suggest an improvement for PBT with at least equivalent toxicity. Some photon-based studies indicate a decrease in testosterone levels after treatment. In contrast,

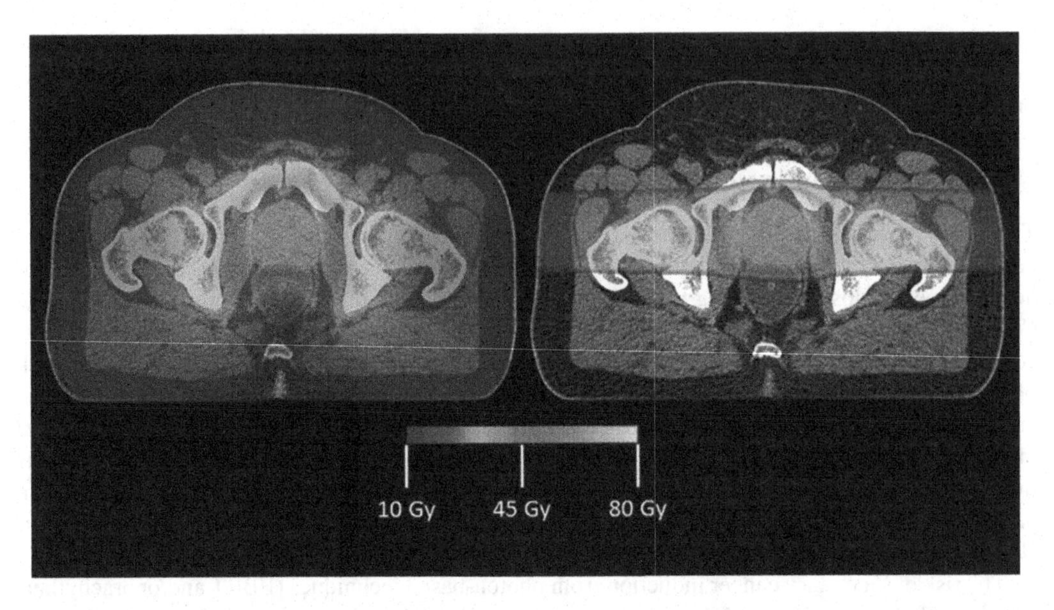

FIGURE 6.3 Examples of typical dose distributions for IMRT (left) and PBT (right) for prostate cancer treatments. Comparison of intensity-modulated radiation therapy (left) vs proton beam therapy (right) for intact prostate cancer. Figure courtesy of Maura Kirk, MS, from the University of Pennsylvania. Image taken from figure in Ojerholm, E., Bekelman, J. 2018. Finding value for protons: the case of prostate cancer. *Seminars in Radiation Oncology* 28:131–137. Published by Elsevier Inc.

several PBT studies show a benefit in preserving post-treatment levels for low- to intermediate-risk patients. Although patient-reported outcome data are still scarce, the signs are also positive for these quality of life (QOL) implications (Wu and Fan 2022; Hoffman 2021).

But there is continuing work for the evidence base comparing the relative toxicity of PBT and IMRT techniques (typical dose distributions are shown in Figure 6.3), using primary endpoints, such as patient-reported bowel function at 2 years, and secondary endpoints, such as urinary and erectile function (Hoffman 2021; Wu and Fan 2022; Poon et al. 2022; Ojerholm and Bekelman 2018). Trials have included PARTIQoL (Prostate Advanced Radiation Technologies Investigating Quality of Life) with a target of 450 men, randomised between PBT and IMRT. Also, the COMPARE study, a parallel cohort study of patients planned for either IMRT or PBT, (with a goal of 1500 patients in each arm), comparing physician-reported bowel toxicity together with freedom from biochemical relapse and QOL with a nest randomisation comparing 60 Gy in 20 fractions and 78Gy in 39 fractions.

Three previous patient-level studies did not show significant, sustained benefit from PBT (Ojerholm and Bekelman 2018). One study suggested PBT patients had lower GU toxicity at 2–3 months, but greater toxicity at one year. The other studies showed no difference in GU side effects (acute or late), although some indications of transient, acute GU toxicity were lower for PBT. There were no differences in sexual side effects.

Within the PRO-PROTON 1 trial, Tilbaek et al. (2023) report on the comparison of photons and protons for prostate and pelvic lymph nodes in high-risk patients. The trial consists of a pilot phase with 40 patients, followed by a randomised controlled trial (RCT) with 400 patients randomised 1:1 between photons and protons. The trial is expected to be completed in three years and will use gold FMs as a surrogate for image guidance. The primary endpoint will be a change (delta score) in patient-reported late GI toxicity at two years compared to baseline. The trial will also compare the delta scores between photon and protons. Secondary endpoints of the trial will be late GU, sexual and GI toxicity, as well as acute GI and GU toxicity. Other endpoints will include health-related quality-of-life scores, biochemical progression-free survival, non-biochemical progression-free survival and overall survival. Follow-up will occur at the end of radiation therapy, four weeks, years 1, 2, 3, 4, 5, 8 and 10, and PROs, QOL and CTCAE data will be collected.

In the study by Forsthoefel et al. (2022), which was a single-institution review of patients treated with histologically confirmed localised prostate cancer. The 33 patients enrolled (July 2018–April 2020) showed no severe acute toxicities. The most common side effect was urinary frequency. With a median follow-up of 18 months, there were no high-grade GU late toxicities, and one Grade 3 GI toxicity. Late erectile dysfunction was found to be common.

In Poon et al.'s (2022) review, disease control was noted. For example, 5 and 8 year biochemical disease-free survival was 5% and 73% respectively for localised prostate cancer; 5 year biochemical recurrence-free survival for intermediate risk after surgical intervention and RT was 85.7%; randomised trials showed similar biochemical failure rates for localised disease comparing high dose PBT and permanent brachytherapy; a single-centre retrospective study showed 5 and 10 year freedom from biochemical relapse rates of 93% and 86% respectively for intermediate risk. Though noted again, there is still currently limited randomised trial results showing efficacy of PBT over photon-based treatments; thus making the data available mixed, with therefore still no overall consensus.

In the recent studies mentioned in Section 6.5.1 above (Choo et al. 2021, 2023; Wong et al. 2022), MPT was used with moderate hypofractionation for prostate and pelvic nodes for high-risk patients. It was a 56-patient prospective study, using IMPT with PBS to deliver doses of 67.5 GyE (45 GyE to the nodes) in 25 fractions simultaneously to CTV high (prostate and SVs) and CTV low (regional pelvic nodes). Late GI and GU toxicities were assessed using CTCAE criteria at baseline, weekly during treatment, three months, then every six months. Generally, treatment was found to be well tolerated with very low rates of Grade 3 GI or GU toxicity. Patients with pre-existing GU symptoms had higher late Grade 2 GU toxicity. Phase III studies are studies needed to assess possible improved therapeutic ratio compared with IMRT (Choo et al. 2021, 2023).

In terms of patient-reported QOL results (using EPIC-26. 26-item Expanded Prostate Index Composite questionnaire), the results acquired at the end of treatment showed a statistically significant and clinically meaningful decline in urinary obstructive (UO) and bowel function (BF) scores (Wong 2022). There was a decline in urinary incontinence (UI) but it was not deemed to be clinically meaningful. UO, UI and BF scores declined at the end of hypofractionated IMPT. UO and UI showed improvement at three months post-treatment and were similar to baseline scores, and BF remained lower at three months post-treatment with a clinically meaningful decline (Wong 2022).

6.8 PROTON SBRT/FOCAL BOOST

Wu and Fan (2022) also discuss PBT SBRT, with studies using moderate (2.4–3.4 Gy per fraction) or ultrahypofractionation (≥ 5 Gy per fraction, delivered in ≤ 5 fractions)(compare with photon methods described in Chapter 8). These are producing extremely promising results in terms of tumour control and toxicity, with the added benefits of greater patient convenience and reduced overall costs.

Recent evidence would suggest that, for low- and intermediate-risk patients, proton SBRT could facilitate effective control with minimal long-term GI and GU toxicities. Some studies, using 36.25 GyE in five fractions, for such patients produced disease-free survival rates of 96.9% and 83.5% for low- and unfavourable intermediate-risk groups, respectively. Favourable long-term rates of 7.8% for Grade 2+ GI toxicity and 5.7% for Grade 2+ GU toxicity were also obtained (Wu and Fan 2022; Hoffman 2021). In comparative PBT studies, comparing 38 GyE in five fractions with standard fractionation of 79.2 GyE in 44 fractions, there were no observable differences in QOoL of patients, with no observable differences in urinary, bowel or sexual function scores from three to 24 months post-treatment and no Grade 3 or higher toxicities for either schedule (Wu and Fan 2022).

Although level I evidence for the role of PBT with SBRT schedules is unclear, available data show promise, often demonstrating comparable efficacy and toxicity when compared with conventional fractionation patterns, with the clear logistical benefits of greater convenience and more cost-effective treatment for patients (Wu and Fan 2022; Hoffman 2021).

In the same way that the development of advanced imaging techniques (such as multi-parametric MR Imaging (mpMRI)) has made focal radiation boost using SBRT-designed image guidance and delivery techniques to the macroscopic visible tumour becoming increasingly viable and feasible for photon treatments, without raising the toxicity profile of patients significantly; one could imagine there is room to develop similar techniques using PBT. Randomised Phase III trials (with nearly 600 patients) for intermediate- to high-risk cancer treated with photon EBRT of 77 Gy (2.2 Gy fractions) compared with additional simultaneous focal boost to the MRI-visible intraprostatic lesion (up to 95 Gy, fractions up to 2.7 Gy) have reported improved biochemical disease-free survival without impacting on toxicity levels and QOL of patients (Wu and Fan 2022).

For PBT, studies have evaluated the feasibility and differences for patients treated with IMPT, compared with VMAT. OAR constraints were not violated, and the superiority of IMPT plans was observed. Newer PET techniques (using prostate-specific membrane antigen (PSMA)-PET) promote the highest degree of accuracy in identifying the focal lesion when combined with mpMRI, and these should improve further both photon and proton treatment techniques for focal boosts (Wu and Fan, 2022), although identifying which patients may benefit most is still a point of debate (Hoffman 2021)

6.9 FUTURES/CARBON IONS

Further improvements in the conformality, accuracy and robustness of PBT for prostate cancer are envisaged through state-of-the-art PBT with intensity modulation (IMPT) with X-ray-based CBCT image-guided and adaptive systems (or even MR-guided facilities) on fully rotational gantry-based PBT delivery systems (Wu and Fan 2022).

But further therapeutic gains are also possible through the use of heavier ions (e.g., carbon ions) compared with PBT, with the dosimetric beam characteristics suggesting a lower-still entry dose and possible further, dosimetric reductions in integral and non-target doses (Bhattacharyya et al. 2019). Prostate case numbers are naturally very small, but where these patients have been treated (usually intermediate-risk cases and with bilateral (alternate days) parallel opposed fields to 52.6 GyE in 12 fractions, using scanning C-ion RT), there have been no cases of Grade 2 or greater skin reactions. No GI reactions were observed, but some Grade 2 GU reactions have been noted. Image guidance was using daily orthogonal X-ray images, compared with DRRs derived from the planning CT scan (Bhattacharyya et al. 2019).

REFERENCES

Albertini, F., Matter, M., Nenoff, L., et al. 2020. Online daily adaptive proton therapy. *British Journal of Radiology*. 93: 20190594.

ASTRO (American Society for Radiation Oncology). 2017. *ASTRO model policies: Proton beam therapy (PBT)*. Chicago, IL: American Medical Association.

Bertholet, J., Knopf, A., Eiben, B., et al. 2019. Real-time intrafraction motion monitoring in external beam radiotherapy. *Physics in Medicine and Biology*. 64: 15TR01.

Bhattacharyya, T., Koto, M., Ikawa, H., et al. 2019. First prospective feasibility study of carbon-ion radiotherapy using compact superconducting rotating gantry. *British Journal of Radiology*. 92: 20190370.

Burnet, N., Mackay, R., Smith, E., et al. 2020. Proton beam therapy: Perspectives on the National Health Service England clinical service and research programme. *British Journal of Radiology*. 93: 20190873.

Choo, R., Hillman, D., Daniels, T., et al. 2021. Proton therapy of prostate and pelvic lymph nodes for high risk prostate cancer: Acute toxicity. *International Journal of Particle Therapy*. 8(2): 41–50.

Choo, R., Hillman, D., Mitchell, C., et al. 2023. Late toxicity of moderately hypofractionated intensity-modulated proton therapy treating the prostate and pelvic lymph nodes for high-risk prostate cancer. *International Journal of Radiation Oncology, Biology, Physics.* 115(5): 1085–1094.

Forsthoefel, M., Hankins, R., Ballew, E., et al. 2022. Prostate cancer treatment with pencil beam proton therapy using rectal spacers sans endorectal balloons. *International Journal of Particle Therapy.* 9(1): 28–41.

Georg, D., Hopfgartner, J., Gòra, J. et al. 2014. Dosimetric considerations to determine the optimal technique for localized prostate cancer among external photon, proton, or carbon-ion therapy and high-dose-rate or low-dose-rate brachytherapy. *International Journal of Radiation Oncology, Biology, Physics.* 88: 715–722.

Hoffmann, A., Oborn, B., Moteabbed, M., et al. 2020. MR-guided proton therapy: A review and a preview. *Radiation Oncology.* 15: 129.

Hoffman, K. 2021. ASTRO 2021: The present and future of proton beam therapy. Presented at ASTRO 2021, 24–27 October 2021, Chicago, IL.

Kirby, M., Calder, K.-A. 2019. *On-treatment verification imaging: a study guide for IGRT.* Boca Raton, FL: CRC Press, Taylor and Francis Group.

Mackay, R. 2018. Image guidance for proton therapy. *Clinical Oncology.* 30: 293–298.

Martins, P. Freitas, H., Tessonier, T., et al. 2021. Towards realtime PGS range monitoring in proton therapy of prostate cancer. *Nature: Scientific Reports.* 11: 15331.

Morrow, N., Lawton, C., Qi, S., et al. 2012. Impact of computed tomography image quality on image-guided radiation therapy based on soft tissue registration. *International Journal of Radiation Oncology, Biology, Physics.* 82(5): 3733–3738. 103: 320–334.

Nogueira, L., Jemal, A., Yabroff, R., et al. 2022. Assessment of proton beam therapy use among patients with newly diagnosed cancer in the US, 2004–2018. *JAMA Network Open.* 5(4): e229025.

Ojerholm, E., Bekelman, J. 2018. Finding value for protons: The case of prostate cancer. *Seminars in Radiation Oncology.* 28: 131–137.

Osman, S., Russell, E., King, R., et al. 2019. Fiducial markers visibility and artefacts in prostate cancer radiotherapy multimodality imaging. *Radiation Oncology.* 14: 237.

Parodi, K. 2020. Latest developments in in-vivo imaging for proton therapy. *British Journal of Radiology.* 93: 20190787.

Poon, D., Wu, S., Ho, L., et al. 2022. Proton therapy for prostate cancer: Challenges and opportunities. *Cancers.* 14: 925.

Reidel, C-A., Horst, F., Schuy, C., et al. 2022. Experimental comparison of fiducial markers used in proton therapy: Study of different imaging modalities and proton fluence perturbations measured with CMOS pixel sensors. *Frontiers in Oncology.* 12: 830080.

Schreuder, A., Shamblin, J. 2020. Proton therapy delivery: What is needed in the next ten years? *British Journal of Radiology.* 93: 20190359.

Scobioala, S., Kittel, C., Wissmann, N., et al. 2016. A treatment planning study comparing tomotherapy, volumetric modulated Arc therapy, sliding window and proton therapy for low-risk prostate carcinoma. *Radiation Oncology.* 11: 128.

Takagi, M., Demizu, Y. Fujii, O., et al. 2021. Proton therapy for localized prostate cancer: long-term results from a single-center experience. *International Journal of Radiation Oncology, Biology, Physics.* 109(4): 964–974.

Tang, S., Both, S., Bentefour, H. et al. 2012. Improvement of prostate treatment by anterior proton fields. *International Journal of Radiation Oncology, Biology, Physics.* 83: 408–418.

Tilbaek, S., Muren, L., Vestergaard, A., et al. 2023. Proton therapy planning and image-guidance strategies within a randomized controlled trial for high-risk prostate cancer. *Clinical and Translational Radiation Oncology.* 41: 100632.

Trofimov, A., Nguyen, P.L., Coen, J.J. et al. 2007. Radiotherapy treatment of early-stage prostate cancer with IMRT and protons: A treatment planning comparison. *International Journal of Radiation oncology, Biology, Physics.* 69: 444–453.

Wong, W., Hillman, D., Daniels, T., et al. 2022. A Phase II prospective study of hypofractionated proton therapy of prostate and pelvic lymph nodes: Acute effects on patient-reported quality of life. *The Prostate.* 82: 1338–1345.

Wu, Y-Y., Fan, K.-H. 2022. Proton therapy for prostate cancer: Current state and future perspectives. *British Journal of Radiology.* 95: 20210670.

7 Margins

7.1 INTRODUCTION

Aspects of margin derivation come from the uncertainties involved in many parts of the radiotherapy process – not least, for image-guided radiotherapy (IGRT) and on-treatment verification imaging, the set-up errors involved interfractionally, internal organ motion and movement intrafractionally, uncertainties in outlining, contouring and the definition of both target and organ at risk (OAR) volumes. Perhaps the most important margin might be regarded as the clinical target volume (CTV)-planning target volume (PTV) margin, but it must not be forgotten that this depends on the assumed accuracy of the CTV itself (McPartlin et al. 2016) and therefore margin reduction must always be exercised with caution. But the ideal is to have as small a margin as possible around the CTV, while (a) assuming the CTV itself is accurate and (b) ensuring the high-dose volume fully encompasses the CTV inter- and intrafractionally, in order to achieve the highest possible tumour control while (through small margins and as small and conformal a high-dose volume as possible) minimising the dose to nearby OARs and non-involved tissues. The smaller the margin, naturally, the greater the likelihood of reducing dose, and therefore toxicity, to non-involved tissues.

In Wang et al.'s (2023) in-depth review, they note that CTV-PTV margins, which were traditionally in the order of 10–15 mm (Dang et al. 2018) to account for unmeasured prostate motion (with all the associated consequences of increased toxicity to normal tissues), can be shrunk with considerable confidence to 7–8 mm with weekly IGRT for prostate cancer (PCa) patients, and down to approximately 5 mm by daily online IGRT, with associated PTV volume reductions then providing the potential to escalate the prescription dose, if deemed wise. Margin reductions have been supported by numerous studies comparing reduced margins with either no image guidance (IG) or IG with standard margins. The clinical outcomes from such studies (specifically in terms of acute genito-urinary (GU) and gastrointestinal (GI) toxicity and late GI toxicity) would support such moves (see Chapters 8 and 9). It must be noted, though, that subsequent dose escalation (because of the confidence from using smaller margins) could then balance out (i.e., negate) such reductions in toxicity, because of the increase in dose beyond the high-dose volume (Wang et al. 2023; Dang et al. 2018).

This is supported in the review by Jackson et al. (2019), indicating that many of the SBRT studies (published between 1990 and 2018) used ≥ 5 mm margins, noting though that magnetic resonance (MR) imaging was not mandated in all the studies, which naturally has a bearing on the margins used. The consequences of margin reductions can be quite striking in terms of PTV volume reduction, with a change from 7 mm down to 3 mm (for typical prostate-only volumes) reduces the overall PTV volume by nearly 40% (Arumugum et al. 2023).

Yartsev and Bauman (2016) examined the issue of target margins explicitly for PCa patients, noting that the margins should ideally be radiotherapy centre- and site-specific, although (at the time) there was no definitive evidence for treatment/IGRT approaches for prostate and therefore no definitive evidence for ideal prostate margins. Other factors, often pragmatic but of real issues, must also be taken into account: e.g., cost-benefit from, say, the introduction of fiducial markers (FMs) or electromagnetic (EM) transponders, with the addition of an invasive procedure for the patient. So, too, one must consider the clinical benefit from, say, increased intrafractional imaging which is deemed necessary, which may incur greater concomitant dose and longer patient in-room time. In more recent times, adaptive procedures have increased staff and machine resource issues (with associated capital and running costs), longer in-room time and therefore reduced throughput (Yartsev and Bauman 2016).

DOI: 10.1201/9781003050988-7

The margins chosen therefore depend on a number of factors, often with the benefit of greater set-up accuracy data acquired through the use of IGRT itself. For example, IGRT on-treatment imaging frequency, modality and timing; the use of daily online versus offline protocols, intrafractional motion estimation from on-treatment imaging before and after treatment delivery, planar vs volumetric (fan-beam and cone-beam) imaging or MV vs kV techniques – all have a bearing on the choice of margin. It also depends upon the clinical protocol and pre-treatment procedures – immobilisation equipment used, target delineation and interobserver variability, treatment intention (e.g., prostate-only versus prostate and/or seminal vesicles (SVs) and/or pelvic lymph nodes (PLNs)) and treatment technique (Yartsev and Bauman 2016; Kirby and Calder 2019, chapter 6).

Within the UK's current national IGRT guidance (On-Target 2) (RCR 2021), it is recommended that data used for margin calculations must be relevant to individual departments; margins should also be consistent with the type of IGRT protocol (see Section 7.2), and volume delineation should be the subject of peer review. A conservative approach is advised for any reduction of margins, because of residual uncertainties, no matter what type of IGRT is used. For prostate treatments specifically, margins can be reduced carefully for prostate treatments, especially using online IGRT, matching to the prostate itself with daily set-up correction. With different IGRT strategies (e.g., bony landmarks or skin marks), margins must be increased, with a likely increase in toxicity (RCR 2021, p. 128).

For more complex treatment arrangements (e.g., involving SVs and PLNs), it must be remembered that motion differs for different parts of anatomy: for prostate, in particular, the movement of prostate, SVs and PLNs is not a rigid body motion (Tudor et al. 2020, p. 14 and p. 50). There can be relative motion between the different structures and caution must be exercised during IGRT with consistent and appropriate use of different anatomical structures to match during on-treatment geometric verification. Delineation uncertainty must also be considered as a factor, it possibly being different for different structures (e.g., for the prostate compared with the SVs) (Tudor et al. 2020, pp. 10–11 and p. 50). One must also consider surrogate error, when the localisation object is not the true clinical target (e.g., bony landmarks instead of the prostate) and, to a lesser extent, migration of FMs with respect to the prostate (Tudor et al. 2020, p. 27 and p. 51).

For the prostate, in particular, pre-treatment preparation and imaging procedures are important and necessary for consistent daily treatments for prostate. Bladder and bowel preparations are recommended, with written guidance for patients (RCR 2021, pp. 127–130), also aiding the use of smaller margins to be applied with greater confidence when used alongside daily online IGRT. In addition, MR scans for treatment planning are desirable, whether used in conjunction with computed tomography (CT) through CT-MR fusion or through MR-only pre-treatment workflows (see Chapter 5). Fusion can be based upon FM positioning or on prostate anatomy itself. MR-informed delineation can be used to help reduce the overall target volume and also the position of any dominant lesion, allowing rational CTV margins (RCR 2021, pp. 127–130).

7.2 APPROACHES FOR IGRT

As mentioned earlier, each IGRT process is associated with different likely residual uncertainties – which must be dealt with through appropriately chosen margins. These should ideally be evaluated and considered within each radiotherapy centre dependent upon their own equipment, protocols used for IGRT for different sites and their own evaluation of set-up errors for their particular techniques and immobilisation methods used (Ghadjar et al. 2019; Dang et al. 2018; Wang et al. 2023).

Key principles include (Ghadjar et al. 2019; Tudor et al. 2020):

- Centres evaluating and estimating their own residual uncertainties and deriving their own particular margins for use for PCa treatments. Factors include whether the IGRT strategy is offline or online and/or uses monitoring/tracking of the target volume or surrogates

- Whether FMs or CT/cone-beam computed tomography (CBCT)-based IGRT approaches are used for accounting for interfractional positional changes
- Whether daily online procedures are preferred for conventional fractionation patterns and recommended for hypofractionated regimes.

Different margins are needed depending on the treatment intention and the target volume definition. As mentioned, in the case of the inclusion of the PLNs or SVs, margins need to be separately applied, compared with the prostate alone. On-treatment set-up based on, say, FMs in the prostate are unlikely to be accurate for motion identified with the SVs and nodes which may each require, therefore, larger margins to ensure the accurate delivery of the prescribed dose throughout the target volume (Ghadjar et al. 2019; Tudor et al. 2020).

Generally speaking, margins (CTV-PTV) are larger for offline-style IGRT approaches (typically of the order of 5–9 mm), compared with daily online approaches (typically 4–6 mm), which themselves are larger than those possible for daily online protocols involving some form of intrafractional monitoring/tracking (typically of the order of 2–4 mm). These are for prostate-only treatments – margins must be larger to account for the greater amplitude and non-uniform movement of larger target volumes (e.g., nodal-involved fields) (Ghadjar et al. 2019).

Fiducial markers are an ideal surrogate for the position of the prostate, with the added benefit of being clearly identifiable on 2D kV planar imaging, but provide little information on internal shape and volume changes (particularly for associated OARs) and also can be unrepresentative of the relative positions of the SVs and PLNs. Volumetric imaging approaches provide more information on shape and relative motion of target volume structures. Common margin recipes can be applied with possible integration of non-rigid residual errors on the SVs and/or PLNs (RCR 2021).

Daily online correction methods using rigid translations are now the most popular IGRT methods for prostate treatments, with set-up correction before the delivery of each fraction. Rotations, non-rigid errors and intrafractional motion can be accounted for, depending on IGRT resources and overall treatment strategies. Smaller margins can be employed compared with offline protocols but, once more, margins are suitably larger and likely necessary for larger target volumes involving the SVs and/or PLNs (Ghadjar et al. 2019).

Real-time monitoring and tracking can be used for prostate-only treatments over few fractions (hypofractionated, stereotactic body radiotherapy (SBRT) treatments), often with technologies which allow for continuous monitoring and correction/tracking of rigid intrafractional motion (see Chapter 5). Residual uncertainties for these cases remain as the intrinsic and residual intrafraction errors (imaging, identification, evaluation and equipment-based). This may depend on the response of the monitoring/tracking systems, changes for treatment delivery systems (e.g., robotic systems (Kirby and Calder 2019, chapter 10) and overall treatment (beam-on) time (Ghadjar et al. 2019; RCR 2021; Tudor et al. 2020).

Yartsev and Bauman (2016) showed clearly how the average values for PTV margins from the literature varied with different IGRT approaches: for external skin marks and room lasers, 10–11 mm; for IGRT with bony anatomy, 8–9 mm; for IGRT with soft-tissue matching, 5-6 mm; for IGRT with fiducial markers, 5–6 mm; for IGRT with EM transponder tracking, 3–4 mm; for IGRT with adaptive radiotherapy (ART), 4–5 mm.

Caution is still needed in the choice of margins noted above (as shown in the error bars on the data in Yartsev and Bauman 2016), since intrafractional motion is known to range from 3 mm with short excursions to greater than 2 cm. For fractionated radiotherapy, these uncertainties and their effect on treatment are likely small, but will be of much greater impact for hypofractionated regimes due to longer beam-on times and less dose averaging over fractions (RCR 2021, p. 129).

The latest British Institute of Radiology (BIR) guidance (Tudor et al. 2020) helps to understand margin requirements when the IGRT is daily and online for all fractions, and examines in depth the nuances and residual uncertainties still present even with such IGRT approaches. Margin calculations and the margins chosen should take the IGRT approach and local technique protocols into account. The report provides useful tools to aid in analysis of geometric uncertainties and also reminds about the validity of the CTV delineation in the first place (Tudor et al. 2020).

Intrafractional motion should also be a consideration for margin calculation, if the IGRT technique does not necessarily consider, monitor or take it into account. Intrafractional motion, even though non-periodic for the prostate (Tudor et al. 2020), could have a bearing on reversing some of the margin gains, with studies using Cine-MR Imaging indicating the magnitude of such motion and also its possible increase with time, with a subsequent need for an increase in PTV margin of perhaps 1–2 mm (Zou et al. 2018).

MR-imaging pre-treatment has the potential to provide smaller contours (from more accurate and confident delineation) and, in the setting of MR-guided ART, margins of 3 mm or less could be used with safety (see Chapters 5 and 8). From the various clinical approaches in development and under trials, clinical results have shown positive outcomes, such as reduced toxicity from margin reductions (see Chapter 9). But, as noted earlier, margin reduction must be judicious, especially when the non-periodic intrafractional motion possible with the prostate is considered and not compensated for (Vargas et al. 2010, Gill et al. 2014; Gill et al. 2015).

7.3 MARGIN CALCULATIONS

The calculation of the CTV-PTV margin is extremely well covered in On-Target 2 (RCR 2021) and in the BIR guidance (Tudor et al. 2020), with worked examples of how systematic and random errors in on-treatment set-ups can be estimated from image-guidance data and then used to calculate the CTV-PTV margin (RCR 2021, pp. 51–61), and how modifications need to be made to account for residual uncertainties present even with daily online IG and set-up correction (Tudor et al. 2020). Key guidance stresses the need for margins to include all elements of uncertainty, while understanding that there is likely always a residual margin (regardless of the IGRT strategy) and that the magnitude of the margin is largely governed by combined systematic errors. Specific examples are included for the case of prostate treatments; training and competency are highlighted and peer review is recommended, especially with regard to delineation uncertainties (RCR 2021; Tudor et al. 2020, pp. 50–53). Authors, such as McNeice et al. (2023), illustrate the use of these guidance documents for margin calculations especially for prostate SBRT-style treatments.

7.4 EXPERIENCES – CONVENTIONAL FRACTIONATION

Within the PACE-B trial (Brand et al. 2019) (details included in Chapters 8 and 9), the conventional/moderately hypofractionated arm of the trial (78 Gy in 39 fractions over 7–8 weeks/62 Gy in 20 fractions over four weeks) had MRI pre-treatment imaging strongly recommended for prostate delineation (through CT-MR fused dataset based upon FMs) (recommended) and recommended a CTV-PTV expansion margin of 5–9 mm isometrically, except for 3–7 mm posteriorly. IGRT was recommended for all patients, with FM (at least three) guidance strongly recommended and accompanied by volumetric tomographic imaging to identify significant changes in rectal position or prostate deformation.

The study of Chaurasia et al. (2018) was a prospective, non-randomised trial using the Calypso EM transponder system for localisation and intrafractional motion monitoring (see Chapter 4) for low- and intermediate-risk prostate patients treated with IMRT (7 or 9 field) on a schedule of 77.4 Gy in 43 fractions (1.8 Gy per fraction). The PTV for all patients was formed from a uniform 3-mm expansion margin around the CTV (with the CTV defined as the GTV without expansion for low-risk patients and prostate, proximal 1-cm of the SVs and a 3-mm expansion of the prostate (minus rectum and bladder) for intermediate-risk patients). With less than 2% of cases having intrafractional movements beyond 3 mm and an acceptable daily treatment time (mean 10 mins, SD 4.8 mins), the group concluded that Calypso tracking was reliable with minimal disruption to daily treatments and an effective way of tracking prostate motion in real time. Dosimetric coverage was adequate with the 3-mm PTV margin and GU and GI toxicity was minimised.

In Arumugam et al.'s (2023) study, data from the SeeDTracker system were used to measure and quantify real-time intrafraction prostate position data. Data were obtained from patients treated using CBCT image guidance for set-up and treated using volumetric modulated arc therapy (VMAT) (with X-ray images acquired at gantry angle spacings of 9 degrees and beam-on times of an average of 2.3 mins (SD 0.3 mins). Treatments were modelled (using Pinnacle Autoplanning) using PTV margins of 6, 5, 4 or 3 mm to a prescription of 60 Gy in 20 fractions, with the prostate deviation data blended into the plans. Standard prostate treatments within the study centre used a 7-mm margin and image guidance using FMs.

Results of the planning study showed that, for patients where intrafractional position correction might be employed, margins of approximately 2 mm were required to meet a 90% population receiving 95% of the prescription dose to the CTV. If position correction was not employed, then margins of 3–4 mm were required to meet the same standard. Notable dose decreases to nearby critical structures were observed. So, the margin reduction was considered feasible when a position monitoring system (like SeedTracker) might be used.

Within the HYPO-RT-PC trial (Widmark et al. 2019) (details included in Chapters 8 and 9), the conventional fractionation arm was 78 Gy in 39 fractions, five days per week over eight weeks. Pre-treatment imaging consisted of CT and T2wMRI; CT-MR fusion for delineation of CTV with the aid of FMs was recommended, but not mandatory, and rectum, anal canal, urinary bladder, penile bulb and femoral heads were contoured as OARs. Margins applied were (for those under the BeamCath IG system) CTV-PTV 6 mm (4 mm posteriorly) for the first four fractions, and, for the remaining fractions (without BeamCath), 10–15 mm margins. For those patients with FMs, treatments were planned with a 7-mm CTV-PTV margin (for both arms of the trial). Treatment delivery was with 3D conformal radiotherapy (CRT), intensity modulated radiotherapy (IMRT) or VMAT, and IG used FMs (orthogonal planer kV imaging) and BeamCath initially for some patients (10%).

7.5 EXPERIENCES – ADAPTIVE/HYPOFRACTIONATION/ INTRAFRACTIONAL MOTION

In the planning study of Kang et al. (2017) used to investigate arc arrangements for prostate SBRT using VMAT, the prostatic bed was delineated as the CTV, with a CTV-PTV margin of 7 mm posteriorly and 10 mm in all other directions. OARS were bladder, rectum and femoral heads. Various arc arrangements were investigated for prescription doses of 42.7 Gy in seven fractions, compared with a standard normally employed of 78 Gy in 39 fractions.

As mentioned above in the PACE-B trial (Brand et al. 2019; see Chapters 8 and 9), the ultra-hypofractionated (UHF) arm was 36.25 Gy in five fractions over 1–2 weeks (daily or on alternate days). For this, the recommended CTV-PTV expansion margin was 4–5 mm isometrically, except for 3–5 mm posteriorly. IGRT was used for all patients, with FM (at least three) guidance strongly recommended, together with tomographic volumetric imaging to identify significant changes in rectal position or prostate deformation. Cyberknife patients had intrafractional motion correction during treatment; for LINAC-based IGRT, most centres permitted the use of VMAT techniques with/without flattening filter (FF). For treatments which were flattening filter free (FFF), beam-on time under three minutes did not warrant intrafraction motion control. However, where beam-on time was greater than three minutes, re-imaging was needed between beams/arcs. Correction shifts ≥ 3 mm were always corrected. Continuous monitoring systems (like Calypso and Clarity ultrasound) were permitted for intrafractional monitoring, with a directive for pauses (and corrections) in treatments if displacement shifts exceeded 3 mm. Gantry-based SBRT using CBCT-based IGRT without fiducials was permitted, with centres demonstrating treatment to the required accuracy (given the significant motion possible which could occur during treatment) to the trial chief investigator (CI) and quality assurance (QA) team.

As mentioned in the HYPO-RT-PC trial (Widmark et al. 2019), one arm was used as an UHF treatment of 42.7 Gy in seven fractions, three days per week over 2.5 weeks inclusive of two weekends. Pre-treatment imaging was as mentioned in Section 7.4 above. Margins were (for BeamCath patients) CTV-PTV of 6 mm (4 mm posteriorly) for all fractions within the UHF arm. Patients with FMs were planned with an isotropic 7-mm CTV-PTV margin for all fractions, with treatment delivery and IGRT as noted in Section 7.4.

Vanhanen et al. (2020) performed a study investigating dosimetric effects of continuous motion monitoring-based localisation (using Calypso EM transponders), gating and intrafractional motion correction for prostate SBRT. Patients were treated with 35 Gy in five fractions (7 Gy per fraction) or 36.25 Gy in five fractions (7.25 Gy per fraction) with treatments on alternate days. Pre-treatment imaging was CT and MR; CT-MR fusion was used for delineation of CTV and OARs, with the CTV defined as the prostate alone and OAR structures were the bladder, rectum and femoral heads.

The CTV-PTV margin was 5 mm (3 mm posteriorly). The urethra was separately delineated as an 8-mm diameter cylindrical structure. Treatment delivery used VMAT with two full or partial arcs. IGRT was FM-based kV imaging using the Calypso EM transponders as fiducials, CBCT and then Calypso EM transponder continuous motion monitoring. During treatment, if prostate motion exceeded 2 mm posteriorly/3 mm elsewhere, irradiation was gated off automatically.

Panizza et al. (2022) undertook a study with patients with organ-confined prostate cancer and dose-escalated SBRT. Patients had bowel (micro-enema) and bladder (100 cm^3 saline) preparation pre-treatment and for each fraction. Pre-treatment imaging consisted of CT and T2wMR Imaging, and CT-MR fusion for CTV and OAR delineation. PTV was the prostate gland and SVs and the treatment schedule was 40 Gy in five fractions or 38 Gy in four fractions consecitively over one week. The planning CTV-PTV margin was a 2-mm isotropic margin, together with a 2-mm margin around the catheter for the EM tracking device, defining a planning organ at risk volume (PRV) for the urethra. OARs were bladder, with PRV of the urethra, rectum, rectum wall and penile bulb.

IGRT was volumetric CBCT with continuous motion monitoring using the RayPilot HypoCath EM transponder device. The RayPilot reference was zeroed after CBCT set up. Treatment delivery was typically two FFF VMAT arcs 10 MV and treatment was interrupted when EM tracking signals exceeded a 2-mm threshold in any direction. Target position was redefined using CBCT after any treatment interruptions.

Among MR-guided studies, Kishan et al. (2023) used an MR LINAC with real-time intrafraction imaging to try to reduce several position uncertainties, more accurate MRI-to-CT fusion and contouring with the use of on-board MRI simulation and better soft-tissue alignment with on-board MR image guidance and intrafraction imaging of the prostate four times per second with automatic beam hold. The MR LINAC was the Viewray MRIdian 0.35T. This trial (MIRAGE) (see also Chapters 8 and 9), was a Phase 3, randomised (between MR-guided and CT-guided image guidance on-treatment) trial of the MRI-Guided Stereotactic Body RT for Prostate Cancer for men with clinically localised prostate cancer at a single cancer centre. Full bladder and empty rectum were used for either arm of the study.

Pre-treatment, fused MR with CT was used for organ delineation. Dose and fractionation were 40 Gy in five fractions, with elective node (25 Gy in five fractions) and simultaneous integrated boost (SIB) to the dominant intraprostatic lesion (DIL) (42 Gy in five fractions) and SIB to the involved pelvic node (35 Gy in five fractions) were allowed at the investigator's discretion. CTV was defined as the prostate and the proximal 1 cm of SV for all patients, with margins of the CT-guided arm CTV-PTV margin 4 mm isotropically. and, for the MR-guided arm, 2 mm isotropically. Treatment delivery was VMAT for the CT-guided arm and step-and-shoot IMRT for the MR-guided arm.

IGRT (see Chapter 8) was for the MR-guided arm, no FMs and 0.35T MR imaging on MR LINAC with prostate alignment for each fraction with a true fast-imaging sequence. During treatment delivery, CINE MRI was used at four frames per second in the sagittal plane. If greater than 10% of the prostate volume was deemed to have moved outside a 3-mm gating boundary around the prostate, the automatic beam hold was initiated.

In the study of Levin-Epstein et al. (2020) to assess optimal PTV margins for prostate cancer, SBRT based on inter- or intrafractional prostate motion was determined from daily image guidance and IG on-treatment data were used to derive optimum margins for SBRT. Data were derived from their standard prostate SBRT protocol as follows.

Pre-treatment processes involved three gold FMs, CT simulation, comfortably full bladder and empty rectum (without specific bowel preparation), with CTV defined as prostate gland only with a mixed use of MR for contouring across the trial centres. Dose was 40 Gy in five fractions. The CTV-PTV margin was 5 mm isotopically, with 3–5 mm posteriorly.

Treatment delivery involved four partial VMAT arcs on alternate days. IGRT's initial set-up used tattoos and lasers or the Exactrac optical system, then kV or MV planar IGRT for FM rigid body matching and CBCT for interfraction motion correction prior to the first half-arc delivery. Repeat planar orthogonal kV images (or Exactrac in-room stereo kV imaging) were used before each subsequent (three) half arcs, to account for intrafractional motion.

From the recorded motion and set-up correction data (inter- and intrafractional data), margins were derived using the van Herk formula. Results showed that a minimum margin of 3 mm was needed to account for intrafractional motion, whereas a minimum of 4 mm was needed when accounting for contouring and other positional uncertainties.

Kuo et al. (2023) published the results reporting their clinical experiences and the feasibility of using 2D-kV planar imaging with online intervention for UHF SBRT for low- to intermediate-risk PCa patients. Pre-treatment procedures were a full bladder and empty rectum (enema), and CT and MR imaging with a normal centre practice being MR simulation for CTV delineation and synthetic CT creation with marker contours. Instead, for this study, CT-MR fusion was used. Planning margins were PTV defined with a 5-mm 3D margin around the prostate gland and bilateral SVs (CTV) except posteriorly, where a 3-mm margin was used. A 2-mm 3D margin was applied to fiducial contours, and hydrogel spacers were used.

Treatment delivery was 15 MV with two VMAT arcs, as 40 Gy in five fractions. IGRT involved three gold FMs with orthogonal kV planar image pairs. Initial set-up was from orthogonal kV image pairs and CBCT images, the latter to ensure proper bladder and rectal volumes, and the relationship to prostate and SVs. During treatment, Intrafraction Motion Review (IMR) software (a system from Varian) was initiated (every 40 degrees, approximately 13 seconds) with manual intervention if two consecutive kV images (approximately 27 seconds apart) detected FM movement beyond the 2-mm threshold. If large excursions (0.5 to 2 cm) were detected, the therapist stopped treatment and reimaged.

Margins derived from monitoring and set-up data included 3D margins, to ensure target coverage, of 3.7 mm, 4.6 mm and 5 mm in left-right (L-R), anterior-posterior (A-P) and superior-inferior (S-I) directions, respectively. Prostate motion occurs over time and SBRT is feasible with online intrafractional monitoring and interventional protocol, here, using 2D kV methods.

Maas et al. (2023) published their study of prostate SBRT with focal SIB, reporting 5-year toxicity, patient-reported quality of life (QOL) outcomes and biochemical recurrence results from this prospective trial with low-/intermediate-risk PCa patients with one or two focal lesions visible on MRI, and an MRI-defined prostate volume within 120 ml. Doses were 40 Gy in five fractions to focal lesion(s) and 36.25 Gy in five fractions to entire prostate within 7–17 days, with delivery of treatment stopped for five consecutive days and treatment on alternate days encouraged. Overall results showed excellent biochemical control without undue late GI or GU toxicity or long-term QOL reduction.

Pre-treatment procedures were 3T MR imaging, with CT simulation with full bladder and empty rectum with CT-MR fusion. CTV1 was the entire prostate gland defined on both CT and MR. The boost volume (CTV2) was the entire volume likely to harbour malignant cells as determined through MRI, clinical examination and biopsy localisation. PTV 1 and 2 were derived from each CTV, using margins of 5 mm in all directions except 3 mm posteriorly, with the same PTV expansion being used for both CTVs.

Treatment delivery was VMAT, with two arcs and 10 MV FFF. IGRT used FMs, with the initial CBCT aligned to FMs and the prostate-rectum boundary, and orthogonal kV radiographs were used to confirm FM positioning. Bladder and rectum volumes were confirmed through CBCT. Intrafraction motion was monitored using triggered kV planar imaging tracking the FMs.

In the recent study of McNiece et al. (2023), intrafraction motion was quantified and subsequent margins were calculated following the latest BIR guidance (Tudor et al. 2020) by daily online IGRT methods in prostate SBRT. Data were collected from 20 patients receiving 36.25 Gy in five fractions (consisting of 97 CBCT image pairs pre- and post-treatment delivery to determine the intrafractional motion). Patients were part of the DELINEATE trial (see Chapters 8 and 9) but were included here in a prospective audit. Pre-treatment procedures were a planned CT simulation, with patients having a comfortably full bladder and empty rectum (using enemas). Planning margins were 6 mm isotropic, 3 mm posteriorly.

Treatment delivery was VMAT with FFF delivery, at 36.25 Gy in five fractions to the prostate and a concomitant boost up to 45 Gy for intraprostatic tumour nodule. IGRT used FMs, an initial CBCT daily and post-treatment CBCT to assess intrafractional movement, with image matching using initial bony landmarks followed by refined matching on FMs. The time between initial CBCT and the start of treatment delivery was recommended to be < 2 minutes, to minimise the effects of intrafractional motion.

Current margins were deemed suitable to cover 92% of recorded intrafractional motion from the study. The results showed that, if the margins were reduced to 3–4 mm, they would encompass 88% of the intrafractional displacements. No changes were considered necessary since clinical outcomes were favourable.

REFERENCES

Arumugum, S., Wong, K., Do, V., et al. 2023. Reducing the margin in prostate radiotherapy: Optimising radiotherapy with a general-purpose linear accelerator using an in-house position monitoring system. *Frontiers in Oncology*. 13: 1116999.

Brand, D., Tree, A., Ostler, P., et al. 2019. Intensity-modulated fractionated radiotherapy versus stereotactic body radiotherapy for prostate cancer (PACE-B): Acute toxicity findings from an international, randomised, open-label, phase 3, non-inferiority trial. *Lancet Oncology*. 20: 1531–1543.

Chaurasia, A., Sun, K., Premo, C. et al. 2018. Evaluating the potential benefit of reduced planning target volume margins for low and intermediate risk patients with prostate cancer using real-time electromagnetic tracking. *Advances in Radiation Oncology*. 3:630–638

Dang, A., Kupelian, P., Cao, M., et al. 2018. Image-guided radiotherapy for prostate cancer. *Translational Andrology and Urology*. 7(3): 308–320.

Dover, L., Dulaney, C. 2023. Prostate stereotactic body radiation therapy margins, accelerated partial breast irradiation techniques, total neoadjuvant therapy local control, hyperfractionated reirradiation, hyaluronic acid rectal spacer, and concurrent docetaxel for head and neck cancer. *Practical Radiation Oncology*. 13(4): 267–272.

Ghadjar, P., Fiorino, C., af Rosenschold, P., et al. 2019. ESTRO ACROP consensus guideline on the use of image guided radiation therapy for localised prostate cancer. *Radiotherapy and Oncology*. 141: 5–13.

Gill, S., Dang, K., Fox, C., et al. 2014. Seminal vesicle intrafraction motion analysed with cinematic magnetic resonance imaging. *Radiation Oncology*. 9: 174.

Gill, S. K., Reddy, K., Campbell, N., Chen, C., Pearson, D. 2015. Determination of optimal PTV margin for patients receiving CBCT-guided prostate IMRT: Comparative analysis based on CBCT dose calculation with four different margins. *Journal of Applied Clinical Medical Physics*. 16(6): 252–262.

Jackson, W., Silva, J., Hartman, H., et al. 2019. Stereotactic body radiation therapy for localized prostate cancer: A systematic review and meta-analysis of over 6000 patients treated on prospective studies. *International Journal of Radiation Oncology, Biology, Physics*. 104(4): 778–789.

Kang, A., Chung, J., Kim, J., et al. 2017. Optimal planning strategy among various arc arrangements for prostate stereotactic body radiotherapy with volumetric modulated arc therapy technique. *Radiology and Oncology*. 51(1): 112–120.

Kirby, M., Calder, K.-A. 2019. *On-treatment verification imaging: A study guide for IGRT*. Boca Raton, FL: CRC Press, Taylor and Francis Group.

Kishan, A., Ma, T., Lamb, J., et al. 2023. Magnetic resonance imaging-guided vs computed tomography-guided stereotactic body radiotherapy for prostate cancer: The MIRAGE randomised clinical trial. *Jama Oncology*. 9(3): 365–373.

Kuo, H., Della-Biancia, C., Damato, A., et al. 2023. Clinical experience and feasibility of using 2D-kVimage online intervention in the ultrafractionated stereotactic radiation treatment of prostate cancer. *Practical Radiation Oncology*. 13: e308–e318.

Levin-Epstein, R., Qiao-Guan, G., Juarez, J., et al. 2020. Clinical assessment of prostate displacement and planning target volume margins for stereotactic body radiotherapy of prostate cancer. *Frontiers in Oncology*. 10: 539.

Maas, J., Dobelbower, M., Yang, E., et al. 2023. Prostate stereotactic body radiation therapy with a focal simultaneous integrated boost: 5-year toxicity and biochemical recurrence results from a prospective trial. *Practical Radiation Oncology*. 13: 466–474.

McNeice, J.M., Sanilkumar, N., Alexander, S.E. et al. 2023. Prostate stereotactic body radiotherapy: quantifying intra-fraction motion and calculating margins using the new BIR geometric uncertainties in daily online IGRT recommendations. *British Journal of Radiology*. 96:20220852.

McPartlin, A., Li, X., Kershaw, L., et al. 2016. MRI-guided prostate adaptive radiotherapy – A systematic review. *Radiotherapy and Oncology*. 119: 371–380.

Morrow, N. Lawton, C., Qi, X., et al. 2012. Impact of computed tomography image quality on image-guided radiation therapy based on soft tissue registration. *International Journal of Radiation Oncology, Biology, Physics*. 82(5): e733–e738.

Panizza, D., Faccenda, V., Lucchini, R., et al. 2022. Intrafraction prostate motion management during dose-escalated linac-based stereotactic body radiation therapy. *Frontiers in Oncology*. 12: 883725.

RCR (Royal College of Radiologists). 2021. *On-target 2: Updated guidance for image-guided radiotherapy*. London, UK: The Royal College of Radiologists.

Tudor, G., Bernstein, D., Riley, S., et al. 2020. *Geometric uncertainties in daily online IGRT: Refining The CTV-PTV margin for contemporary photon radiotherapy*. London: The British Institute of Radiology.

Vanhanen, A., Poulsen, P., Kapanen, M. 2020. Dosimetric effect of intrafraction motion and different localization strategies in prostate SBRT. *Physica Medica*. 75: 58–68.

Vargas, C., Saito, A., Hsi, W., et al. 2010. Cine-magnetic resonance imaging assessment of intrafraction motion for prostate cancer patients supine or prone with and without a rectal balloon. *American Journal of Clinical Oncology*. 33(1): 11–16.

Wang, S., Tang, W., Luo, H., et al. 2023. The role of image-guided radiotherapy in prostate cancer: A systematic review and meta-analysis. *Clinical and Translational Radiation Oncology*. 38: 81–89.

Widmark, A., Gunnlaugsson, A., Beckman, L., et al. 2019. Ultra-hypofractionated versus conventionally fractionated radiotherapy for prostate cancer: 5-year outcomes for the HYPO-RT-PC randomised, non-inferiority, phase 3 trial. *Lancet*. 394: 385–395.

Yartsev, S., Bauman, G. 2016. Target margins in radiotherapy of prostate cancer. *British Journal of Radiology*. 89: 20160312.

Zou. W., Dong, L., Teo, B.-K. 2018. Current state of image guidance in radiation oncology: Implications for PTV margin expansion and adaptive therapy. *Seminars in Radiation Oncology*. 28: 238–247.

8 Hypofractionated Regimes

Localised prostate cancer (PCa) patients have traditionally been treated with conventional fractionation schemes of, e.g., 78 Gy in 39 fractions. In the past decade or so, data has shown that moderate hypofractionation (treatment with about 20 fractions) is not inferior to conventional schemes, a finding supported by evidence, for PCa has a low α/β ratio and therefore lends itself to treatment with greater than 2 Gy fractions, making moderate hypofractionation (fractional doses of 2.5 Gy and above) a recommendation (alongside conventional schemes), provided that treatment is delivered with intensity modulated radiotherapy (IMRT) or volumetric modulated arc therapy (VMAT) and with good image-guided radiotherapy (IGRT) on-treatment verification (RCR 2019). As techniques, equipment and IGRT itself have improved over the past decade too, stereotactic body radiotherapy (SBRT) has evolved, harnessing these technical developments and enabling the clinical testing of schedules of just five fractions – what some would describe as ultrahypofractionated (UHF) regimes (Tree et al. 2022).

For such schedules, with highly conformal dose distributions and tight planning margins, IGRT is crucial to successful clinical implementation and brings the possibility of treating with even smaller margins (see Chapter 7). As pre-treatment imaging has improved, through the availability of positron emission tomography-computed tomography (PET-CT) imaging and, more recently, multiparametric magnetic resonance (MR) imaging (mpMRI), techniques have also developed to identify patients with a dominant intraprostatic lesion (DIL) and perform a simultaneous integrated boost (SIB) to that focal lesion. Combined with the possibilities of online MR-guided adaptive radiotherapy (see Chapter 5) and more methods to understand/monitor/account for intrafractional motion (see Chapter 5), IGRT has become the mainstay for enabling these UHF to be successfully tried, with clinical trial results now forthcoming and outcomes being published (see Chapter 9). The results are such that treatments below five fractions are now being trialled, potentially exploiting the low α/β ratio still further to gain a better therapeutic index.

This chapter examines a sample of recently published work, trial results and techniques on these hypofractionated and UHF trials for PCa patients.

Within the UK, the main clinical guidelines for dose and fractionation are published by the Royal College of Radiologists (RCR 2019). Within that guidance, hypofractionation is defined as doses of 2.5 Gy and above per fraction. Quoted too are key references from the published literature for the evidence behind historical trial data for the recommendations made. A number of historical and randomised studies using what might be described as moderate hypofractionation of 16–20 fractions (of 2.5 Gy and above) have demonstrated a non-inferiority to, and toxicity as low as, conventional (e.g., 39-fraction) fractionation.

Profound extreme or ultra-hypofractionation (within this chapter from this point described as ultrahypofractionated (UHF) regimes) may be defined as fraction sizes of 6 Gy and above (RCR 2019). These have been shown to be feasible and safe in cohort studies, with high disease control for low-risk PCa patients. The PACE and HYPO-RT-PC trials (see Table 8.1 below) are two such trials quoted within the RCR guidelines, representing regimes of 36.25 Gy in five fractions and 42.7 Gy in seven fractions, respectively.

Wolf et al. (2021) have published a statement from the German Society for Radiation Oncology (DEGRO) on UHF for prostate cancer. As extreme or ultra-hypofractionation techniques are commonly synonymous with stereotactic body radiotherapy/stereotactic ablative radiotherapy (SBRT/SABR), they recommend the definition of UHF as being fraction sizes of 4 Gy and above; and 2.2–4 Gy per fraction as moderate hypofractionation. This avoids the gap in definitions which is present with ASTRO/ASCO/AUA guidance, which defines moderate hypofractionation as 2.4–3.4 Gy per

DOI: 10.1201/9781003050988-8

TABLE 8.1
The Key Details for Some of the Hypofractionated and UHF SBRT Techniques Discussed in this chapter for PCa Patients, Including Trials with a Focal Boost Treated with a Simultaneous Integrated Boost.

Study	Pre-treatment Procedures	Planning Details	Dose (Gy)	Fraction Number	IGRT	Intrafractional Motion	Comments
Maas 2023	TRUS-guided three gold FMs; mpMRI/spacers not required. CT simulation with full bladder and empty rectum; CT/MR fusion. CTV1 entire prostate gland defined on CT and MRI; boost (CTV2) volume most likely to have malignant cells as determined on MR, clinical exam and biopsy; PTV1 and 2 created with 5 mm margins (except 3 mm posteriorly) on CTVs respectively	VMAT, two arcs 10 MV FFF photons	36.25 (40 to focal lesion)	Five (over 7–17 calendar days; five consecutive days prohibited; delivery Every other day (EOD) encouraged but not mandatory)	CBCT aligned to FMs and prostate/rectum border; bladder/rectum status assessed; orthogonal kV images on FMs;	Monitored with triggered kV imaging on FMs	PCa SBRT SIB prospective trial; low-/intermediate- risk; one or more focal lesions on MRI; 26 patients; median follow-up 59.5 months
Murray 2020; Tree 2023	TRUS- or TPUS- guided three gold FMs; mpMRI and CT scans; moderately full bladder and rectal preparation (micro-enema); knee and ankle immobilisation; fused CT-MR to aid outlining; PTV boost mpMRI-defined boost plus 2 mm margin excluding urethra; margins varied from 0 mm posteriorly/3 mm elsewhere for highest dose PTV; isotropic 6 mm for lowest dose PTV	IMRT (5 or 7 field, SnS IMRT)	74 (82 to focal lesion) 60 (67 to focal lesion)	37 20	Daily, online FMs with rectal/bladder filling assessment	Not specified in publication	DELINEATE; single centre, prospective Phase 2 multicohort study comparing standard (Cohort A) and hypofractionated (Cohort B) dose escalation to intraprostatic tumour nodule in localised PCa; SIB; 55 patients (Cohort A); 158 patients (Cohort B); median follow-up 74.5 months (Cohort A), 52.0 months (Cohort B); for 2023 paper, > 60 months

(Continued)

TABLE 8.1 (CONTINUED)
The Key Details for Some of the Hypofractionated and UHF SBRT Techniques Discussed in this chapter for PCa Patients, Including Trials with a Focal Boost Treated with a Simultaneous Integrated Boost.

Study	Pre-treatment Procedures	Planning Details	Dose (Gy)	Fraction Number	IGRT	Intrafractional Motion	Comments
Draulans 2020; De Cock 2023	TPUS- or TRUS- guided three or four FMs; CT and mpMR imaging (mpMR used to identify the DIL); CT/MR fusion; comfortably full bladder, empty rectum (rectal balloon used in one centre); CTV – whole prostate + 4 mm margin around mpMR visible tumour nodule (GTV); CTV-PTV margin 4–5 mm	VMAT (dual), > 6MV	35/ 50 (simultaneous integrated focal boost)	Five weekly fractions over 29 days Five bi-weekly over 15 days for second trial (2023 paper)	Daily online with either orthogonal kV planar or CBCT	Not specified in publication	Hypo-FLAME; intermediate-/ high- risk PCa patients on a prospective Phase 2 trial; treated with extreme hypofractionated doses of 35 Gy in five weekly fractions with a 50 Gy SIB focal boost; 100 patients in one weekly trial; 124 patients in bi-weekly trial (2023 paper)
Alayed 2020	TP FMs; supine, vac-lock bag; comfortably full bladder, empty rectum; CT and MR; CTV1 – nodal CTV and SVs; CTV2 – prostate only; 6-mm margin around CTV1, 3-mm margin around CTV2 to generate respective PTVs; mpMRI detected DIL	Not specified in publication	40 prostate/25 pelvis 35 prostate/25 pelvis/50 focal boost	Five Five	Daily online CBCT pre and post treatment	Not specified in publication	Two Phase 2 trials of prostate and pelvic SABR, with or without SIB to the DIL; high-risk patients with 40 Gy to the prostate/25 Gy to pelvis in five fractions; intermediate- and high-risk patients with 35 Gy to the prostate/25 Gy to the pelvis/50 Gy SIB to the DIL all in five fractions; 30 patients to each trial; follow-up > 24 months

(Continued)

TABLE 8.1 (CONTINUED)

The Key Details for Some of the Hypofractionated and UHF SBRT Techniques Discussed in this chapter for PCa Patients, Including Trials with a Focal Boost Treated with a Simultaneous Integrated Boost.

Study	Pre-treatment Procedures	Planning Details	Dose (Gy)	Fraction Number	IGRT	Intrafractional Motion	Comments
Hannan 2022	FMs; peri-rectal hydrogel spacer; full bladder, empty rectum; CT simulation; mpMRI; 3-mm margin around prostate and proximal 1-cm SV; 0–3 mm around DIL for focal boost; SB Frame for immobilisation, vac-lock bag	Dynamic Arc IMRT (Linac) delivering SABR with 6 degree of freedom treatment couch	47.5 prostate/50 focal boost/22.5 to PLNs Dose escalation by cohort to 52.5 and 55 for focal boost; 25 for PLNs	Five	Daily CBCT; FMs surrogate for prostate, pelvic bony anatomy surrogate for PLN field; bladder and rectum status checked on CBCT	Not specified in publication	Phase 1 prospective trial, high-risk PCa, well-defined prostatic lesion on mpMRI; Four dose escalation cohorts with 7–15 patients in each, performed sequentially and 90 day observation period
Ong 2023	TRUS-guided FMs; planning CT, mpMRI to identify DIL; full bladder and empty rectum; GU-lock and endorectal immobilisation; CT/MR fusion; GTV – mpMRI defined DIL; CTV prostate gland +/- 1 cm expansion into base of SVs; PTV – CTV plus 2-mm margin AP and Lateral, 2.5-mm SI	VMAT; hexapod; 6MV on CT-based LINAC	26 (32 to GTV (focal boost)	Two (one week apart)	Daily online CBCT pre and post treatment (CT-based Linac)	Not specified in publication	2SMART Phase 2, single arm prospective trial, safety of two-fraction SABR with focal boost to MR defined DIL for localised PCa patients; low-/ intermediate-risk patients

(*Continued*)

TABLE 8.1 (CONTINUED)
The Key Details for Some of the Hypofractionated and UHF SBRT Techniques Discussed in this chapter for PCa Patients, Including Trials with a Focal Boost Treated with a Simultaneous Integrated Boost.

Study	Pre-treatment Procedures	Planning Details	Dose (Gy)	Fraction Number	IGRT	Intrafractional Motion	Comments
Westley 2023	MR imaging, mpMRI used to identify DIL; CT simulation; contours and reference plan generated on MRI acquired on the MR LINAC; full bladder, empty rectum (micro-enemas); GTV – DIL (for 2 fraction arm); CTV1 – prostate + 1 cm SVs; 3-mm CTV-PTV margin	11-field IMRT on MR LINAC; target and OARs edited daily and new plans created for each fraction; two fraction treatments delivered in two sequential subfractions; average treatment times – 59 mins (five fraction regime) and 140 mins (two fraction regime)	36.25 24 (27 Focal boost to DIL)	Five (EOD excluding weekends, over 2 weeks) Two (7 days apart, over 8 days)	Adapt-to-Shape workflow; imaging immediately before beam on and non-negligible displacement corrected using Adapt-to-position.	Patients on two fraction regime, partial emptying of bladder and 20 min wait before second subfraction – readjustment of plan if necessary	HERMES – single centre, noncomparative randomised Phase 2 trial for intermediate- or (lower) high-risk PCa patients; 1:1 allocation to either five fraction or two fraction treatments with a SIB to the MR defined DIL for the two-fraction arm only; treatment delivered on an MR LINAC (Elekta UNITY) with daily online adaption; intermediate analysis after ten patients treated in each group and reporting 12 weeks post treatment
Acklin-Wehnert 2023	Alpha cradle or Combi-fix immobilisation, supine; full bladder, empty rectum; CTV – prostate and involved SV(s); 5-mm margins in all directions or 8-mm (5-mm posteriorly); MR/CT imaging	IMRT or VMAT	70 or 60 (SV involvement) 58.8 or 50 (no SV involvement)	28 or 20 (SV involvement) 28 or 20 (no SV involvement)	Daily CBCT and/or alignment to FMs	Not specified in publication	Intermediate- or high-risk patients treated with moderate hypofractionation of PCa with SV involvement evident on MR or clinical exam; 41 patients treated and propensity score matched with 82 treated in the same period to prostate alone; 26% had SV involvement, 95% were high-risk

(Continued)

TABLE 8.1 (CONTINUED)
The Key Details for Some of the Hypofractionated and UHF SBRT Techniques Discussed in this chapter for PCa Patients, Including Trials with a Focal Boost Treated with a Simultaneous Integrated Boost.

Study	Pre-treatment Procedures	Planning Details	Dose (Gy)	Fraction Number	IGRT	Intrafractional Motion	Comments
Kishan 2023	CT simulation, FMs for CT patients; no FMs for MR patients; free-breathing fast 0.35T MR simulation scan; full bladder, empty rectum; fused 1.5T MR scans used to aid contouring (for both arms); CTV prostate plus proximal 1-cm SVs; CT – margin 4-mm; MR – margin 2-mm; Elective nodal RT (25 Gy), SIB for DIL (42 Gy) and SIB for PLN if involved (35 Gy) permitted at investigator discretion	Median post-imaging delivery times – CT 232 secs (3.9 mins); MR 1133 (19 mins); CT delivery on C-arm LINAC (VMAT), MR on Viewray MRIdian LINAC (SnS IMRT)	40	Five (EOD)	CT-arm; CBCT + ortho kV planar on FMs MR-arm, true fast MR imaging	CT-arm; none MR-arm; cine MRI four times per second in sagittal plane; if > 10% prostate volume outside 3-mm gating boundary, automatic beam hold initiated.	MIRAGE; Non-blinded, single-centre, Phase 3 randomised trial comparing MR guidance with CT guidance (conventional, control arm) for SBRT PCa patients; 1:1 randomisation; 77 patients in CT arm, 79 patients in MR
Lukka 2018; Lukka 2023	Not specified in publications	Not specified in publications	36.25 or 51.6 Cyberknife, IMRT/VMAT or PBT	Five or 12, respectively Five – over 2 weeks and one day, twice a week; 12 – over 2.5 weeks, 5 times a week	Required but not specified in publications	Not specified in publications	NRG Oncology RTOG 0938; two ultra-hypofractionated trials; nonblinded, randomised Phase 2 study; low-risk patients in each arm (five or 12 fractions) compared with historical control; 127 patients (five fraction arm); 128 patients (12 fraction arm); median follow-up for all patients was 3.8 years (45.6 months – 2018 paper); 5.38 years (64.6 months – 2023 paper)

(Continued)

TABLE 8.1 (CONTINUED)
The Key Details for Some of the Hypofractionated and UHF SBRT Techniques Discussed in this chapter for PCa Patients, Including Trials with a Focal Boost Treated with a Simultaneous Integrated Boost.

Study	Pre-treatment Procedures	Planning Details	Dose (Gy)	Fraction Number	IGRT	Intrafractional Motion	Comments
Widmark 2019; Fransson 2021	Gold FMs; CT and T2w MR imaging; CTV delineated on CT (SVs not treated), co-registration with MR recommended but not mandatory; FM patients – CTV-PTV 7-mm isotropic margin for all patients (both groups)	3D-CRT (10 MV) or IMRT (6 MV or higher)	42.7 78	Seven – three days per week (EOD) for 2.5 weeks 39 – five days per week for eight weeks	BeamCath (Bergstrom et al. 1998) for first 10% of patients; Gold FMs with daily online, orthogonal kV or MV portal imaging or CBCT using FMs	Not specified in publications	HYPO-RT-PC; open-label, randomised, Phase 3 non-inferiority trial for intermediate-to-high-risk PCa patients; randomised to 42.7 Gy in seven fractions or 78 Gy in 39 fractions; 1,200 patients assigned (602 conventional; 598 ultra-hypofractionated); median follow-up of 5.0 years (IQR 3.1-7.0) (60 months – 2019 paper); 4.0 years (IQR 2.1-6) (48 months – 2021 paper; patient -reported QoL outcomes)
Tree 2022; Van As 2023	Three or four FMs recommended but not mandated; bowel prep (enema) suggested together with moderate bladder filling; no rectal spacer; CT and MR (strongly recommended) fused for outlining; CTV – prostate only (for low-risk) or prostate plus proximal 1-cm SVs (for intermediate risk); for CRT CTV-PTV margin 5–9-mm isometric (3–7-mm posterior); for SBRT margin 4–5-mm isometric (3–5-mm posterior)	Not specified in publications	36.25 78 or 60	Five over 1–2 weeks (daily or EOD at centre discretion) 39 or 20 over 7–8 or 4 weeks respectively	Daily (FMs or CBCT) was mandatory	For SBRT, continuous intrafractional motion monitoring was permitted; or re-imaging required if fraction delivery exceeded 3 min	PACE-B; open-label, multicohort, randomised, controlled Phase 3 trial for low-to-intermediate risk histologically-confirmed PCa; randomised to SBRT 36.25 in five fractions or CRT 78 Gy in 39 fractions or (protocol amendment) 62 Gy in 20 fractions; 874 patients (441 CRT, 433 SBRT); 24 months reporting (2022 paper); median follow-up 73.1 months (IQR 62.6-84.0) (2023 paper)

(Continued)

TABLE 8.1 (CONTINUED)

The Key Details for Some of the Hypofractionated and UHF SBRT Techniques Discussed in this chapter for PCa Patients, Including Trials with a Focal Boost Treated with a Simultaneous Integrated Boost.

Study	Pre-treatment Procedures	Planning Details	Dose (Gy)	Fraction Number	IGRT	Intrafractional Motion	Comments
Gorovets 2020	TRUS-guided three FMs; full bladder, empty rectum; CT and 3T MR scans; CT/MR fusion; MR-only simulation for some patients; supine with custom thermoplastic mould; CTV – prostate plus SVs; PTV 5-mm circumferential/3-mm posterior margins	SBRT technique – two VMAT arcs, 15 MV	40 (median); 45; 37.5; 42	Five fractions Five; five or six fractions, respectively	Orthogonal kV imaging with FMs and CBCT to confirm bladder and rectal volumes; subsequent orthogonal kV image pair	Motion tracking during VMAT delivery using simultaneous MV-kV orthogonal imaging; treatments interrupted for motion > 1.5–2 mm	SBRT patients treated using an MV-kV motion management system; 193 patients; median dose 40 Gy in five fractions (159 patients); 45 Gy in five (n=5); 37.5 Gy in five (n=9); 42 Gy in six (n=10)

Some abbreviations: CTV – clinical target volume; PTV – planning target volume; FM – fiducial marker; PCa – prostate cancer; TRUS – transrectal ultrasound; TPUS – transperineal ultrasound; DIL – dominant intraprostatic lesion; SnS – step-and-shoot; FFF – flattening filter free; SVs – semnal vesicles; PLNs – pelvic lymph nodes; EOD – every other day; PBT – proton beam therapy; mpMRI – multiparametric magnetic resonance imaging; SBRT – stereotactic body radiotherapy fraction, and ultrahypofractionated regimes as those with a dose per fraction of 5 Gy or higher. This naturally leaves an undefined zone between 3.4 and 5 Gy per fraction.

Wolf et al. (2021) emphasise that UHF regimes should usually be delivered using high-precision techniques (LINAC or robot-based) aided by daily image guidance and including some form of adequate motion management intrafractionally to allow for smaller margins and high-dose conformation.

Intrafractional motion management is stressed in On-Target 2 (RCR 2021), in that it is known to range from 3 mm with small deviations as much as 22 mm for the prostate, making the impact normally small for fractionated therapy, but possibly greater in hypofractionated (and especially UHF) regimes where the impact and risk associated with intrafractional motion can be greater especially with longer beam-on time and less averaging across fractions. It is therefore important to be efficient but accurate for image registration and correction of on-treatment set-up errors and consider repeat imaging (after the fraction has been delivered) and/or intrafractional correction mid-way through the fraction for ultra-hypofractionated regimes. This is especially important for online adaptive radiotherapy (ART) (see Chapter 5).

In Ghadjar et al.'s (2019) ESTRO ACROP consensus guidelines, note is made of several randomised studies and much current practice underlining the efficacy of moderately hypofractionated treatment regimes for PCa. But there is still limited experience and clinical outcome data for UHF regimes, with fraction sizes in the range of about 7–8 Gy. Often, such techniques make use of implanted fiducial markers (FMs) for real-time monitoring and/or tracking on standard LINAC or robotic delivery platforms, because of the greater issue with intrafractional motion for these highly conformal, high dose per fraction, small-margin treatments. Ultrahypofractionation has been shown to be feasible and well tolerated, with many studies now presenting outcome data beyond 5–7 years (see Chapter 9). Margins are naturally smaller (about 3 mm; see Chapter 7) and studies are showing high rates of biochemical control and low rates of severe toxicities with longer follow-ups of 5–7 years (Kishan et al. 2019).

In terms of ASTRO/ASCO/AUA guidelines for clinically localised PCa (Eastham et al. 2021), Guideline 30 indicates that clinicians should offer moderate hypofractionated external beam radiotherapy (EBRT) for patients with low- or intermediate-risk PCa who elect for EBRT. This is noted as a strong recommendation, with Grade A level evidence. Guideline 31 indicates that clinicians may offer ultrahypofractionated EBRT for patients with low- or intermediate-risk PCa who elect for EBRT, but notes this as a conditional recommendation with Grade B level evidence. Eastham et al. (2021) quote the review undertaken by Hickey in 2019, which included ten randomised trials examining > 2.5 Gy per fraction regimes, for which the pooled analysis demonstrated no differences in biochemical recurrence-free, metastasis-free, PCa-specific survival, or overall survival when compared with conventional fractionation. Additionally, there were no differences in acute genito-urinary (GU) or late gastrointestinal (GI) toxicity from conventional treatments.

Eastham et al. (2021) also quote the HYPO-RT-PC trials (Widmark et al. 2019; Fransson et al. 2021) (see Table 8.1) ultrahypofractionated trial of 42.7 Gy in seven fractions (6.1 Gy per fraction), which was compared with conventional 78 Gy in 39 fractions for image-guided techniques, using 3D-conformal radiotherapy (3D-CRT), IMRT or VMAT for intermediate- or high-risk localised PCa. The UHF regime was found to be non-inferior to the conventional scheme for failure-free survival, PCa mortality and overall survival. Although associated with a slightly higher incidence of acute urinary and bowel symptoms, no differences were identified in terms of late symptoms or quality of life (QoL).

From Loblaw's (2020) Canadian research, the general argument is that UHF regimes should be considered a standard of care for intermediate-risk PCa patients. The definition of 5 Gy per fraction or higher is used for UHF, and it is put forward also that those men with low-risk PCa, who decline active surveillance and choose EBRT, should be offered UHF radiotherapy as an alternative to conventional fractionation. For those with intermediate risk, UHF may be offered as an alternative, but it is strongly encouraged to be part of a clinical trial or multi-institutional registry. Much debate is drawn as to whether UHF should now be more of a standard of care option for such patients.

Loblaw (2020) quotes from three randomised controlled trials (RCTs) testing moderate hypofractionation for non-inferiority to conventional treatments (Catton 2007; Dearnaley et al. 2016; Lee et al. 2016) and four RCTs testing whether moderate hypofractionation is superior to conventional treatments (Aluwini 2015; Pollack 2013; Hoffman 2018; Arcangeli 2017). All had median follow-up over five years and showed non-inferiority and negative superiority. UHF studies (such as HYPO-RT-PC and PACE-B (see Table 8.1 and Chapter 9)) are showing positive results; e.g., for the HYPO-RT-PC trial, no differences in failure-free survival and no significant differences in acute or late Grade 2+ GI or GU toxicities were reported, although some patient reported outcomes (PROs) for the acute GI and GU domains did show slightly poorer results which generally settle over time. Positive results have been most recently published for the PACE-B trial (Tree et al. 2022; Van As et al. 2023).

So, even with older technology and techniques, UHF regimes show significant promise in terms of non-inferiority and toxicity; some acute toxicities may be reduced further with a lengthened overall treatment time (e.g., by treating every other day (EOD)), making UHF regimes even better tolerated. With research into UHF having been conducted for over 20 years, there is a strong case for its further adoption, possibly as new standards of care.

In terms of SBRT focal boost treatments, Zhao et al. (2023) have published a very recent review of *in-silico* studies and clinical trials. For some cases of PCa, there is an expected elevated tumour control without an accompanying increase in toxicity if one delivers an escalated dose to a biologically defined sub-volume within the target volume; a focal boost. A number of both *in-silico* studies and clinical trials are now in the evidence base for the incorporation of such dose escalation to the dominant intraprostatic lesion (DIL), usually as simultaneous integrated boosts (SIBs). For the studies (nearly 70) reviewed by Zhao and colleagues (2023), there was a consistent reporting of significant dose escalation to the gross tumour volume (GTV) and higher tumour control; there was little or no significant difference in toxicity between focal boost and conventional RT. Most prevalent were acute ≥ Grade 2 GU toxicities, while the least prevalent were late ≥ Grade 2 GI. There was a negative correlation between the level of toxicity and the proportion of low- or intermediate-risk patients in the cohorts. An overall conclusion was drawn that focal boosting has the potential to be a new standard of care for these patients, with (on average) boost doses to GTVs for the clinical trials at about 126% of the prescribed dose to the remaining prostate. Beyond those using more conventional regimes, ultra-hypofractionated SIB studies included reports by Alayed et al. (2020), Herrera et al. (2019), Hannan et al. (2022), Draulans et al. (2020) and Marvaso et al. (2020) (see Table 8.1 and Chapter 9), some with treatment to the prostate and focal boost to the DIL, and some with simultaneous multiple dose levels to prostate (+/- seminal vesicles (SVs)), DIL and pelvic lymph nodes (PLNs). Most regimes are of five fractions.

In Kishan et al.'s (2019) study, pooled analysis of individual patient data of cohort studies of low- and intermediate-risk PCa patients (2,142 in total) treated with SBRT, 7-year incidence of biochemical recurrence was found to be 4.5% and 10.2% for low- and intermediate-risk patients, respectively, with the 7-year cumulative incidence of severe GU toxicities being 2.4% and 0.4% for severe GI toxicities, respectively. Of the 2142 patients included, 1185 were classed as low risk, 692 as intermediate (favourable), and 265 as intermediate (unfavourable); median follow-up was 6.9 years.

For the 12 prospective studies examined in Kishan et al.'s (2019) work, ten used a delivery in five fractions and two used four fractions; fraction size varied from 6.7 to 9.5 Gy, with approximately 47% of patients receiving daily fractions, 47% EOD and 5% weekly. Delivery ranged from non-coplanar (robotic gantry) to non-opposing coplanar fields to step-and-shoot (SnS) IMRT and VMAT, with margins ranging from 2 to 5 mm from the prostate (0–5 mm posteriorly); all used FMs with either orthogonal imaging (four) prior to treatment, additional orthogonal imaging during treatment (one) or real-time tracking of FMs during delivery (eight).

In essence, the pooled analysis showed that SBRT for low- and intermediate-risk disease resulted in low rates of severe toxicities and high rates of biochemical control, suggesting that these ultrahypofractionated regimes are an appropriate, definitive treatment for such PCa patients.

Within Jackson et al.'s (2019) systematic review of patients on prospective SBRT studies, 38 trials and prospective series were identified treating a total of 6,116 patients, the mean number of fractions was five (4–9) and the mean dose per fraction at the trial level was 7.25 Gy (5–10 Gy); most patients were low or intermediate risk (45% and 47%, respectively)

In terms of MR guidance, Kishan et al. (2023) have recently published their latest results from the MIRAGE trial comparing MR-guided and CT-guided SBRT for PCa patients. Clinical trial data continue to demonstrate the effectiveness of SBRT regimes for PCa patients. MR guidance is used for contour delineation, as well as specific imaging (mpMRI) for identifying DILs for patients where a focal boost (delivered simultaneously and integrated with the main treatment) may be advantageous. MR guidance on-treatment has now developed with the advent of MRLINACs (see Chapter 5; and Kirby and Calder 2019, chapter 10), offering even greater advantages to PCa patients.

Theoretical advantages for MR guidance on-treatment (through MRLINACs) include direct prostate motion monitoring (rather than through surrogates, such as FMs); improved soft-tissue contrast for helping with set-up accuracy through improved image matching on-treatment; reduction in residual errors present in CT/MR fusion procedures through the direct use of MR images for contouring; MR-MR fusion for contouring (say, with diagnostic MR) and for registration with on-treatment MR (where image matching is MR to MR). Together, these aspects could enable even further margin reduction and potential reductions in OAR/normal tissue toxicity. The Kishan et al. (2023) study (MIRAGE trial) is a randomised trial comparing MR-guided and CT-guided (as the conventional, control arm) SBRT treatments for PCa patients, some of which included those requiring a focal boost to the DIL.

Finally, the Westley et al. (2023) study (HERMES trial) is another MR-guided trial. It is a randomised trial comparing MR-guided SBRT treatments, but also with the utilisation of ART strategies for PCa patients requiring a focal boost to the DIL, and comparing two and five fraction regimes. Adapt-to-shape workflows (see Chapter 5) are used; with target and OARs edited daily and new plans created each fraction and repeat MR images acquired immediately prior to beam-on with non-negligible shifts in target anatomy being corrected using adapt-to-position protocols (see Chapter 5). This is done for both arms of the study.

REFERENCES

Acklin-Wehnert, S., Carpenter, D., Natesan, D., et al. 2023. Toxicity and outcomes of moderately hypofractionated radiation for prostate cancer with seminal vesicle involvement. *Advances in Radiation Oncology*. 8: 101252.

Alayed, Y., Davidson, M., Liu, S., et al. 2020. Evaluating the tolerability of a simultaneous focal boost to the gross tumor in prostate SABR: A toxicity and quality-of-life comparison of two prospective trials. *International Journal of Radiation Oncology, Biology, Physics*. 107: 136–142.

Aluwini, S., Pos, F., Schimmel, E., et al. 2015. Hypofractionated versus conventionally fractionated radiotherapy for patients with prostate cancer (HYPRO): Acute toxicity results from a randomised noninferiority phase 3 trial. *Lancet Oncology*. 16: 274–283.

Arcangeli, G., Saracino, B., Arcangeli, S., et al. 2017. Moderate hypofractionation in high-risk, organ confined prostate cancer: Final results of a phase III randomized trial. *Journal of Clinical Oncology*. 35: 1891–1897.

Bergstrom, P., Lofroth, P-O., Widmark, A. 1998. High-precision conformal radiotherapy (HPCRT) of prostate cancer – a new technique for exact positioning of the prostate at the time of treatment. *International Journal of Radiation Oncology, Biology, Physics*. 42(2):305–311.

Catton, C., Lukka, H., Gu, C-S., et al. 2007. Randomized trial of a hypofractionated radiation regimen for the treatment of localized prostate cancer. *Journal of Clinical Oncology*. 35: 1884–1890.

De Cock, L. Draulans, C., Pos, F., et al. 2023. From once-weekly to semi-weekly whole prostate gland stereotactic radiotherapy with focal boosting: Primary endpoint analysis of the multicentre phase II hypo-FLAME 2.0 trial. *Radiotherapy and Oncology*. 185: 109713.

Dearnaley, D., Syndikus, I., Mossop, H., et al. 2016. Conventional versus hypofractionated high-dose intensity-modulated radiotherapy for prostate cancer: 5-year outcomes of the randomised, non-inferiority, phase 3 CHHiP trial. *Lancet Oncology*. 17: 1047–1060.

Draulans, C., van der Heide, U., Haustermans, K., et al. 2020. Primary endpoint analysis of the multicentre phase II hypo-FLAME trial for intermediate and high risk prostate cancer. *Radiotherapy and Oncology*. 147: 92–98.

Eastham, J. Auffenberg, G., Barocas, D., et al. 2021. Clinically localised prostate cancer: AUA/ASTRO guideline. Part III: principles of radiation and future directions. *The Journal of Urology*. 208: 26–33.

Fransson, P., Nilsson, P., Gunnlaugsson, A., et al. 2021. Ultra-hypofractionated versus conventionally fractionated radiotherapy for prostate cancer (HYPO-RTP-PC): Patient-reported quality-of-life outcomes of a randomised, controlled, non-inferiority, phase 3 trial. *Lancet Oncology*. 22: 235–245.

Ghadjar, P., Fiorino, C., af Rosenschold, P., et al. 2019. ESTRO ACROP consensus guideline on the use of image guided radiation therapy for localised prostate cancer. *Radiotherapy and Oncology*. 141: 5–13.

Gorovets, D., Burleson, S., Jacobs, L., et al. 2020. Prostate SBRT with intrafraction motion management using a novel linear accelerator-based MV-kV imaging method. *Practical Radiation Oncology*. 10: e388–e396.

Hannan, R., Salamekh, S., Desai, N., et al. 2022. SABR for high-risk prostate cancer: A prospective multilevel MRI-based dose escalation trial. *International Journal of Radiation Oncology, Biology, Physics*. 113: 290–301.

Herrera, F., Valerio, M., Berthold, D., et al. 2019. 50-Gy stereotactic body radiation therapy to the dominant intraprostatic nodule: Results from a phase 1a/b trial. *International Journal of Radiation Oncology, Biology, Physics*. 103: 320–334.

Hoffman, K., Voong, K., Levy, L., et al. 2018. Randomized trial of hypofractionated, dose-escalated, intensity-modulated radiation therapy (IMRT) versus conventionally fractionated IMRT for localized prostate cancer. *Journal of Clinical Oncology*. 36: 2943–2949.

Jackson, W., Silva, J., Hartman, H., et al. 2019. Stereotactic body radiation therapy for localised prostate cancer: A systematic review and meta-analysis of over 6,000 patients treated on prospective studies. *International Journal of Radiation Oncology, Biology, Physics*. 104(4): 778–789.

Kirby, M., Calder, K-A. 2019. *On-treatment verification imaging: A study guide for IGRT*. Boca Raton, FL: CRC Press, Taylor & Francis Group.

Kishan, A., Dang, A., Katz, A., et al. 2019. Long-term outcomes of stereotactice body radiotherapy for low-risk and intermediate-risk prostate cancer. *JAMA Network Open*. 2(2): e188006.

Kishan, A., Ma, T., Lamb, J., et al. 2023. Magnetic resonance imaging-guided vs computed tomography-guided stereotactic body radiotherapy for prostate cancer: The MIRAGE randomised clinical trial. *JAMA Oncology*. 9(3): 365–373.

Lee, W., Dignam, J., Amin, M., et al. 2016. Randomized phase III noninferiority study comparing two radiotherapy fractionation schedules in patients with low-risk prostate cancer. *Journal of Clinical Oncology*. 34: 2325–2332.

Loblaw, A. 2020. Ultrahypofractionation should be a standard of care option for intermediate-risk prostate cancer. *Clinical Oncology*. 32: 170–174.

Lukka, H., Deshmukh, S., Bruner, D., et al. 2023. Fiver-year patient-reported outcomes in NRG oncology RTOG 0938, evaluating two ultrahypofractionated regimens for prostate cancer. *International Journal of Radiation Oncology, Biology, Physics*. 116(4): 770–778.

Lukka, H., Pugh, S., Bruner, D., et al. 2018. Patient reported outcomes in NRG oncology RTOG 0938, evaluating two ultrahypofractionated regimens for prostate cancer. *International journal of Radiation Oncology, Biology, Physics*. 102(2): 287–295.

Maas, J., Dobelbower, M., Yang, E., et al. 2023. Prostate stereotactic body radiation therapy with a focal simultaneous integrated boost: 5-year toxicity and biochemical recurrence results from a prospective trial. *Practical Radiation Oncology*. 13: 466–474.

Marvaso, G., Gugliandolo, S., Bellerba, F., et al. 2020. Phase II prospective trial "Give Me Five" short-term high precision radiotherapy for early prostate cancer with simultaneous boost to the dominant intraprostatic lesion: The impact of toxicity on quality of life (AIRC IG-13218). *Medical Oncology*. 37: 74.

Murray, J., Tree, A., Alexander, E., et al. 2020. Standard and hypofractionated dose escalation to intraprostatic tumor nodules in localised prostate cancer: Efficacy and toxicity in the DELINEATE trial. *International Journal of Radiation Oncology, Biology, Physics*. 106(4): 715–724.

Ong, W., Cheung, P., Chung, H., et al. 2023. Two-fraction stereotactic ablative radiotherapy with simultaneous boost to MIR-defined dominant intra-prostatic lesion – Results from the 2SMART phase 2 trial. *Radiotherapy and Oncology*. 181: 109503.

Pollack, A., Walker, G., Horwitz, E., et al. 2013. Randomized trial of hypofractionated external-beam radiotherapy for prostate cancer. *Journal of Clinical Oncology*. 31: 3860–3868.

RCR (Royal College of Radiologists). 2019. *Radiotherapy dose fractionation*, 3rd edition. London: The Royal College of Radiologists.

RCR (Royal College of Radiologists). 2021. *On-target 2: updated guidance for image-guided radiotherapy*. London: The Royal College of Radiologists.

Tocco, B., Kishan, A., Ma, T., et al. 2020. MR-guided radiotherapy for prostate cancer. *Frontiers in Oncology*. 10: 616291.

Tree, A., Ostler, P., van der Voet, H., et al. 2022. Intensity-modulated radiotherapy versus stereotactic body radiotherapy for prostate cancer (PACE-B): 2-year toxicity results from an open-label, randomised, phase 3, non-inferiority trial. *Lancet Oncology*. 23: 1308–1320.

Tree, A., Satchwell, L., Alexander, E., et al. 2023. Standard and hypofractionated dose escalation to intraprostatic tumor nodules in localised prostate cancer: 5-year efficacy and toxicity in the DELINEATE trial. *International Journal of Radiation Oncology, Biology, Physics*. 115(2): 305–316.

Van As, N., Tree, A., Patel, J., et al. 2023. 5-year outcomes from PACE-B: An international phase III randomised controlled trial comparing stereotactic body radiotherapy (SBRT) vs. conventionally fractionated or moderately hypo fractionated external beam radiotherapy for localised prostate cancer. *International Journal of Radiation Oncology, Biology, Physics*. 117(4): e2–e3.

Westley, R., Biscombe, K., Dunlop, A. et al. 2024. Interim toxicity analysis from the randomised HERMES trial of 2- and 5-fraction magnetic resonance imaging-guided adaptive prostate radiation therapy. *International Journal of Radiation Oncology, Biology, Physics*. 118(3):682–687.

Widmark, A., Gunnlaugsson, A., Beckman, L., et al. 2019. Ultra-hypofractionated versus conventionally fractionated radiotherapy for prostate cancer: 5-year outcomes for the HYPO-RT-PC randomised, non-inferiority, phase 3 trial. *Lancet*. 394: 385–395.

Wolf, F., Sedlmayer, F., Aebersold, D., et al. 2021. Ultrahypofractionation of localised prostate cancer. *Strahlentherapie und Onkologie*. 197: 89–96.

Zhao, Y. Haworth, A., Rowshanfarzad, P., et al. 2023. Focal boost in prostate cancer radiotherapy: A review of planning studies and clinical trials. *Cancers*. 15: 4888.

9 Clinical Outcomes

The ultimate result one must look for following developments in treatment planning, on-treatment verification, online adaptive processes and accurate treatment delivery is that of the clinical outcomes – how our radiotherapy techniques fare as far as tumour control and normal tissue toxicities are concerned. *In-silico* studies can help to simulate the likely clinical gains from, for example, tighter planning margins, improved image guidance or intrafractional controls, but the clinical results are what indicate success for our patients.

In this chapter, we examine some recently published evidence examining results from conventionally fractionated treatments and from ultrahypofractionated (UHF) and focal boost techniques and early MR-guided adaptive studies for prostate cancer (PCa) patients.

Perhaps the most comprehensive and recent systematic review of results from image-guided treatments for PCa patients is that by Wang et al. (2023). As well as ensuring that the conformal high-dose volume is delivered as planned to the patient's anatomy, image-guided radiotherapy (IGRT) also helps to inform our treatment methods, especially with respect to the clinical target volume to planning target volume (CTV-PTV) margin, to ensure that the high-dose volume is likely to always cover the CTV, and also make sure that the doses to organs at risk (OARs) are within applied constraints and always as low as achievable (Wang et al. 2023).

From weekly IGRT and especially with the advent of online, daily IGRT, studies have shown that it is possible to reduce margins from what might be 7–15 mm (and now lower), thereby reducing the PTV volume and providing the opportunities to escalate prescription doses safely. To achieve this, some studies have demonstrated a mitigation of biochemical recurrence rates by up to 14% by cone-beam computed tomography (CBCT) image guidances (Wang et al. 2023).

But IGRT is not all positive – some reports do not demonstrate improvements in biochemical failure-free survival (BFFS) with an escalation of prescription dose; some even report worse overall survival (OS) for daily IGRT when compared with weekly on-treatment verification results, the assumption being that perhaps there is an increase in secondary cancer mortality (SCM) involved with the higher frequency of on-treatment imaging (Wang et al. 2023). More frequent IGRT can be associated with increases in overall treatment times (which can affect overall throughput and treatment room workload) and, in some healthcare settings, an increased financial burden. The additional (concomitant) dose may be an issue for some treatments, with clinical consequences in terms of late effects or secondary cancer induction (Wang et al. 2023).

Wang et al.'s (2023) review reports on studies on men with nonmetastatic PCa, treated using any form of IGRT. A total of 18 studies, published between 2009 and 2021, were included for final review and meta-analysis. By far the majority (17/18; 94%) used some form of daily IGRT; and there was an approximately 50:50 split in studies which used standard or reduced CTV-PTV margins. Most used 1.8–2 Gy fraction sizes; three of the studies included patients treated on a higher dose per fraction (2.7–3.65 Gy), which might be defined as ultrahypofractionated. Approximately 12 of the studies made use of fiducial markers (FMs) with orthogonal or CBCT imaging for IGRT.

In broad terms, the results of the meta-analysis showed:

- IGRT significantly reduced acute genito-urinary (GU), acute gastrointestinal (GI) and late GI toxicity compared with non-IGRT treatments
- Daily IGRT demonstrated an improvement in three-year prostate-specific antigen (PSA) replacement-free survival (PRFS) and biochemical failure-free survival (BFFS), compared with non-daily IGRT

DOI: 10.1201/9781003050988-9

- No significant effects of IGRT on late GU toxicity, five-year overall survival (OS) or secondary cancer mortality (SCM) were found
- 2D imaging with FMs as a surrogate could be seen as being more beneficial for late GU toxicity than other types of IGRT
- High-frequency daily IGRT could lead to greater three-year biochemical failure-free survival (BFFS) benefits for prostate patients than weekly IGRT
- IGRT and treatment with reduced margins could significantly reduce acute GU toxicities, but escalating prescription dose might affect this decrease in acute GU toxicity
- In technology combinations, IGRT combined with IMRT might be more effective than 3D-CRT for protecting against acute GU and rectal toxicities.

Limitations were identified by several authors within the review (Wang et al. 2023); some study results may be variable in quality; some comparisons may be weaker because of the limited numbers of studies (e.g., only two publications for the IGRT frequency comparators); most studies featured 3D-conformal radiotherapy (CRT), so the effects of the now more widely used intensity modulated radiotherapy/volumetric modulated arc therapy (IMRT/VMAT) techniques may affect toxicity results in the future, while the 5-year follow-up analysis may be underpowered for some outcome indicators. The majority of publications were retrospective studies, and therefore more higher-quality RCTs are needed to further verify the results and the role of IGRT in current radiotherapy and IGRT techniques (including MR-guided adaptive radiotherapy (MRgART) and CBCT-based ART) as well as trials associated with UHF schedules.

In Ghadjar et al.'s (2019) ESTRO ACROP consensus guidelines, daily online corrections are recommended over off-line procedures because of the proven benefit of biochemical progression-free survival (bPFS) and rectal toxicity. Monitoring and ideally tracking of intrafractional motion may be considered for extreme hypofractionation (see Chapter 5), although its clinical relevance has not yet been established.

Within the guidelines, it is noted that clinical outcomes depend not only upon treatment techniques and PTV margins, but also on the definition of the CTV. Recent (2018) ESTRO ACROP guidelines on computed tomography (CT)- and magnetic resonance (MR)-based target volume delineation give recommendations for CTV delineation, including expansion of the prostate contour to allow for potential extracapsular extension which may be needed, particularly for intermediate- and high-risk patients (Ghadjar et al. 2019).

In some published results not covered in Wang et al.'s (2023) review, it was found, for 962 patients, treated with a CT-based offline adaptive IGRT technique and a median follow-up of 5.5 years, that rectal volume on the initial planning CT scan was predictive of neither bPFS nor toxicity. Elsewhere, IGRT with FMs has been found to be a risk factor for biochemical failure with a median follow-up of 53 months. However, for that study, only 25 patients were included and lateral margins were 3 mm, with the possible conclusion that extensive margin reduction was likely detrimental in that case and was to be avoided; this shows that margin reduction beyond what is considered necessary for CTV expansion should be avoided; and also illustrates the criticality of clear CTV definition, especially if small CTV-PTV margins are employed (Ghadjar et al. 2019).

At the time of Ghadjar's (2019) publication, moderate hypofractionation is noted as showing outcomes equivalent to conventional fractionation patterns, using the same delivery techniques. But limited results were available for UHF regimes with doses of 7–8 Gy per fraction. Many studies were found to employ FMs for real-time monitoring and/or tracking with standard LINAC or robotic treatment platforms, as intrafractional motion is more relevant in these schedules (see Chapter 8). But UHF regimes were proving to be feasible and well tolerated, albeit within limited follow-up. One cohort study (2,142 men in 12 Phase 2 SBRT trials) showed that most centres used a 3-mm posterior margin, but only two of the 12 studies didn't use intrafractional tracking/motion management, opting instead for wider, 4–5-mm posterior margins (see Chapter 7). With a median follow-up of seven years, high rates of biochemical control and low rates of severe toxicities were reported.

As noted in Chapter 8, the move towards UHF regimes for localised PCa patients is growing, as early and on-going clinical trial results are published. From Loblaw's (2020) Canadian group, comparisons were made of over 600 patients across five major centres in Canada. The patient group was low-risk PCa treated with a stereotactic ablative radiotherapy (SABR) technique (35 Gy, five fractions over 29 days) and compared with conventional low doserate (LDR) brachytherapy and external beam radiotherapy (EBRT) (74–79.8 Gy, 37–42 fractions over 7.5–8.5 weeks). There were no significant differences found in biochemical disease-free survival (bDFS) before or after matching SABR against LDR brachytherapy with a median follow-up of over 50 months (Loblaw 2020).

Studies using SABR on robotic and C-arm LINACs of over 2,000 patients with low and intermediate risk, combined across ten single-institutional and two multi-institutional prospective Phase 2 trials with a median follow-up of over 83 months, produced a seven-year biochemical failure (recurrence-free survival) rate of 10.2% for intermediate-risk and 4.5% for low-risk patients. Crude incidence of Grade 3 and GU and GI toxicities were 0.6% and 0.1%, respectively) (Loblaw 2020; Kishan et al. 2019; Wolf et al. 2021).

Loblaw (2020) also quotes from three randomised controlled trials (RCTs) testing moderate hypofractionation for noninferiority to conventional treatments (Catton 2007; Dearnaley et al. 2016; Lee et al. 2016) and four RCTs testing whether moderate hypofractionation is superior to conventional (Aluwini 2015; Pollack 2013; Hoffman 2018; Arcangeli 2017). All had median follow-up over five years and showed noninferiority and negative superiority, respectively. UHF studies (such as HYPO-RT-PC and PACE-B (see Table 8.1 and Chapter 9)) show positive results, e.g., for the HYPO-RT-PC trial, no difference in failure-free survival and no significant differences in acute or late Grade 2+ GI or GU toxicities were observed, although some patient reported outcomes (PROs) for the acute GI and GU domains did show slightly worse results which generally settled with time. Positive results have been most recently published for the PACE-B trial (Tree et al. 2022; Van As et al. 2023).

One might note that the PACE-B trial had a shorter overall treatment time (OTT) compared with HYPO-RT-PC (1.5 weeks compared with 2.5 weeks, respectively) but still showed slightly lower toxicity, which could be attributed to image guidance (see Chapter 8, Table 8.1) and the tighter margins (4–5 mm (3–5 mm posteriorly) compared with 7 mm for HYPO-RT-PC). OTT is likely to have an effect on acute toxicity; some of the UHF five-fraction studies ranged from 7 to 29 days in terms of treatment time, representing treatments within a week, every other day (EOD) and through to once weekly. Some results show superior acute bowel quality of life (QoL) and moderate-to-severe problems for once weekly delivery compared with EOD, without compromising medium-term outcomes like biochemical failure or metastasis rates (Loblaw 2020).

So, even with what one may describe as older technology and techniques, UHF regimes show significant promise in terms of noninferiority and toxicity; some acute toxicities may be reduced further with a lengthened overall treatment time (e.g., by treating EOD), further improving the tolerance of UHF regimes. With research into UHF having been conducted for over 20 years, there are strong cases for its further adoption, possibly as new standards of care (Loblaw 2020).

From Wolf et al.'s (2021) published statement from the German Society for Radiation Oncology (DEGRO) on UHF for treatment of prostate cancer, the HYPO-RT-PC trial (see Chapter 8, Table 8.1) is of note, in which randomised men with intermediate- to high-risk PCa, received either 42.7 Gy in seven fractions, three days per week, EOD over 2.5 weeks or a conventional 78 Gy in 39 fractions, five days per week over eight weeks. Reported at the time was a median follow-up of five years, and a failure-free survival equivalent (84%) in both arms. Acute Grade 2 or worse GU toxicity was slightly increased in the UHF arm at the end of treatment, but rates were equivalent at five years. No difference in GI toxicity was noted at any timepoint.

Wolf (2021) notes some interesting differences from other studies: low-risk patients were excluded, MR was not used for contouring, and LINAC-based treatments were not always IMRT

– 80% were 3D-CRT. Furthermore, PTV margins were 7 mm, large compared with studies like the PACE-B trial.

Alongside the statement's review of HYPO-RT-PC, the PACE-B, noninferiority trial for men at low or intermediate risk of PCa is randomised to receive either 36.25 Gy in five fractions over 1–2 weeks or conventional/moderate hypofractionated regimens (78 Gy in 39 fractions over 7–8 weeks or 62 Gy in 20 fractions over four weeks). There was no significant difference found in acute toxicity, with a slight favouring of the SBRT regime, this finding contrasting with some of the results in the HYPO-RT-PC trial (see later in this section).

In Jackson et al.'s (2019) review of 38 prospective studies, covering over 6,000 patients, where most were low- (45%) or intermediate-risk (47%) categories, they concluded overall that UHF schedules could be considered a standard strategy for localised PCa, although other commentators reserve judgement until more high-risk patients are included in such studies.

Wolf and colleagues (2021) conclude that retrospective and randomised prospective data with follow-up beyond five years now show outcome results comparable to moderate hypofractionation trials for both biochemical control and toxicity profiles. Further results are needed from UHF regimes for high-risk patients, but from centres experienced with SBRT, with high technical standards (e.g., MR-based planning; IMRT/VMAT treatment delivery; daily IGRT with basic intrafractional control (like imaging after 3 minutes treatment time) to safely accompany PTV margins around 3 mm; short treatment times (through VMAT and/or flattening filter free (FFF) techniques); max fraction sizes of 8 Gy; follow-up of not less than five years and recommended registry inclusion). Through these measures, these regimes could seem to be justified for low- and intermediate-risk PCa patients even outside clinical trials (Wolf et al. 2021).

Already mentioned is the HYPO-RT-PC trial, for which the main publications are Widmark et al. (2019) and Fransson et al. (2021) (see Chapter 8, Table 8.1). The 2019 paper published the five-year outcomes of the randomised, non-inferiority, Phase 3 trial; where 1,200 patients were randomised (approx. 600 in each arm), consisting of 89% intermediate-risk and 11% with high-risk disease. Median follow-up was 5.0 years.

Results showed a failure-free survival of 84% in both arms, with a slight increase in frequency of acute, physician-reported Grade 2 or worse GU toxicity for the UHF group at the end of treatment. No significant differences were observed in Grade 2 or worse GU or GI toxicity after RT at five years. Patient-reported outcomes (PROs) showed significantly higher acute GU and GI symptoms for the UHF arm, but no difference in late symptoms. Overall, though, the UHF treatment was deemed non-inferior for intermediate- to high-risk patients regarding failure-free survival. Early side effects were more pronounced for the UHF regime, whereas late toxicity was similar for both treatment arms. Results supported the use of UHF regimes for these PCa patients (Widmark et al. 2019).

The 2021 paper (Fransson et al. 2021) published PROs (quality of life (QoL)) over a long-term (> five years (median 48 months; IQR 25–72)). PROs on QOL show the UHF regime was as well tolerated as conventional fractionation – with no statistically significant differences in proportions of patients with clinically relevant acute urinary symptoms or problems, and sexual function between two treatment groups at the end of radiotherapy. There were also no clinically relevant differences in urinary, bowel, or sexual functioning thereafter – again increasing the support for the use of this type of regime for intermediate to high-risk patients.

Also already mentioned is the PACE-B trial, for which the main publications are Tree et al. (2022) and Van As et al. (2023) (see Chapter 8, Table 8.1). Within the earlier paper, two-year toxicity results were reported from this open-label, randomised, Phase 3, non-inferiority trial, which involved 874 men (430 in the conventional arm and 414 in the SBRT arm). Toxicity profiles showed a two-year RTOG Grade 2 or worse GU toxicity to be 2% for the conventional arm and 3% for the SBRT arm. For two-year RTOG Grade 2 or worse GI toxicity, the values were 3% for the conventional arm and 2% for the SBRT arm. No serious, adverse events (Grade 4 or worse) or treatment-related deaths were recorded and, overall, the toxicity rates were deemed to be similar, making the SBRT regime a safe one, associated with low rates of side effects.

In the very recently reported results for the 2023 paper, 5-year outcomes were published, indicating a primary endpoint of freedom from biochemical/clinical failure (BCF) after a median follow-up of 73.1 months (IQR 62.6–84.0). The five-year BCF event-free rate was 94.6% for the conventional arm and 95.7% for the SBRT arm. The five-year RTOG Grade 2 or worse GU toxicity was 3.2% and 5.5% for the conventional and SBRT arms, respectively. The five-year RTOG Grade 2 or worse GI toxicity was down to 0.3% for the conventional schedule and 0.3% for the SBRT one. Overall BCF event-free rates were high for all PACE-B participants; five-fraction SBRT was, therefore, considered to be non-inferior to the conventional treatment for BCF. With reduced patient attendances and shorter treatment times, the authors felt that the five-fraction SBRT regime should be the new standard of care for low-/favourable intermediate-risk-localised PCa patients (Van As et al. 2023).

Two key publications are also in circulation for the American NRG Oncology RTOG 0938 trial, namely Lukka et al. (2018) and Lukka et al. (2023). The latest results report on five-year PROs in the evaluation of two UHF regimes for PCa. Non-blinded, randomised, Phase 2 studies of low-risk patients were compared with historic controls, with the UHF regime being 36.25 Gy in five fractions over 2 weeks and 1 day, twice a week treatments (7.25 Gy per fraction), compared with 51.6 Gy in 12 fractions over 2.5 weeks, five times a week treatments (4.3 Gy per fraction). Based on changes in bowel and urinary domains and toxicity (early and late), the two SBRT regimes were well tolerated. Disease-free survival (DFS) was 89.6% in the five-fraction regime and 92.3% in the 12-fraction regime, with no late Grade 4 or 5 treatment-related GU or GI toxicities noted.

Outcomes are also being reported for studies where a simultaneous integrated boost (SIB) (focal) is being used, for which the review by Zhao et al. (2023) is perhaps the most recent, examining both *in-silico* study results and clinical data. As noted, for some cases of PCa, there is an expected elevated tumour control without an accompanying increase in toxicity if one delivers an increased dose to a biologically defined sub-volume within the target volume – a focal boost. A number of studies suggest, therefore, the incorporation of such dose escalation to the dominant intraprostatic lesion (DIL), usually as SIBs.

For the (nearly 70) studies reviewed by Zhao and colleagues (in the clinical trials covering mostly intermediate- (41%) or high- (47%) risk patients), there was a consistent reporting of significant dose escalation to the GTV and higher tumour control; there was little or no significant difference in toxicity between focal boost and conventional RT (Zhao et al. 2023). Most prevalent were acute ≥ Grade 2 GU toxicities; least prevalent were late ≥ Grade 2 GI toxicities There was a negative correlation between the rate of toxicity and the proportion of low- or intermediate-risk patients in the cohorts. An overall conclusion was drawn that focal boosting has the potential to be a new standard of care for these patients.

Some 79.4% of the trial studies used multiparametric MR imaging (mpMRI) to identify the DIL, whereas the others used either positron emission tomography (PET), PET-CT, mpMRI or other methods. Most studies (21/34, 62%) used IMRT or VMAT for a treatment modality, and 9/27 (33%) of the trial studies did not apply a margin around the GTV for the focal boost. From the trial studies, for Grade 2 or worse toxicities, the rates across the trials were (1) Acute GU – approximately 33%; (2) Acute GI – approximately 14%; (3) Late GU – approximately 19%; and (4) Late GI – approximately 11%. On average, the boost doses to GTVs for the clinical trials were about 126% of the prescribed dose to the remaining prostate.

In a slightly more complex approach, Hannan et al. (2022) reported on SABR used for high-risk PCa patients with a prospective multi-level dose escalation strategy (see Chapter 8, Table 8.1), delivering dose levels simultaneously integrated into the five-fraction treatment, to the prostate, the mpMRI-defined DIL and to the pelvic lymph nodes (PLNs). From the study, the rates for Grade 2 or worse toxicities were (1) Acute GU – approximately 25%; (2) Acute GI – approximately 13%; (3) Late GU – approximately 20%; and (4) Late GI – approximately 7%. The two-year actuarial BC control rates were reported as 96.6%, biochemicali progression-free survival rate was 94.8%, disease-specific survival rate was 100% and overall survival rate was 98.2% (Hannan et al. 2022).

Also of note is the Spanish study by Zapetero et al. (2022), using an MRI-guided focal boost to DIL for this Phase 2 trial of 30 patients. Unlike the UHF regimes, dose and fractionation were 76 Gy in 35 fractions, with a concomitant focal boost to the MR-defined DIL to 85 Gy using VMAT and FM-based IGRT, withdaily online CBCT. Primary endpoints were biochemical disease-free survival, mpMRI-defined local control, and acute toxicity, and median follow-up was 30 months (IQR 26–40).

Results showed that all patients remained free of biochemical relapse. Complete response of the mpMRI-defined DIL was observed in 25 patients at six months, and at 9 months for the remaining five patients; 20% of the patients experienced acute Grade 2 GU toxicity but none had Grade 3, whereas acute GI toxicity was reported in 7% of patients (only Grade 1). No Grade 2 or worse GU or GI late toxicity were observed. From this, the overall conclusions were that the schedule produced excellent morphological and functional response control, with a safe toxicity profile.

Also including a focal boost is the DELINEATE trial (Murray et al. 2020; Tree et al. 2023) (see Chapter 8, Table 8.1). Again, a more conventional fractionation schedule was used here, but with an additional cohort using a moderately hypofractionated regime. The DELINEATE trial is a single-centre, prospective, Phase 2 multicohort study: Cohort A; 74 Gy in 37 fractions; Cohort B 60 Gy in 20 fractions). IGRT and IMRT are used for PCa patients with intermediate- and high-risk disease. Simultaneous integrated boosts were used in each cohort, with 82 Gy in Cohort A and 67 Gy in Cohort B to lesions visible under mpMRI investigations, with 55 patients treated in Cohort A and 158 in Cohort B.

In the 2020 paper, the median follow-up was at approximately 75 months for Cohort A and 52 months for Cohort B. For the 2023 paper, the median follow-up was at > 5 years for each cohort. From the study, the rates for Grade 2 or worse toxicities were: (1) one-year GU – approximately 0% (Cohort A), 10% (Cohort B); (2) one-year GI – approximately 4% (Cohort A), 8% (Cohort B); (3) five-year GU – approximately 13% (Cohort A), 18% (Cohort B); and (4) five-year GI – approximately 13% (Cohort A), 15% (Cohort B). Five-year freedom from biochemical/clinical failure was approximately 98% for Cohort A and 97% for Cohort B. Results together indicate a safety, tolerance and feasibility for focal boosting in both conventional and moderately hypofractionated regimes with a low chance of recurrence at five years (Murray et al. 2020; Tree et al. 2023).

Draulans et al. (2020) reported on the hypo-FLAME trial (see Chapter 8, Table 8.1), a prospective Phase 2 trial for intermediate- or high-risk PCa patients, all treated with an extreme hypofractionated regime of 35 Gy in five weekly fractions to the prostate (OTT 29 days), with an integrated boost to 50 Gy for the mpMRI-identified tumour. A total of 100 patients were enrolled, treated across four academic centres, and all were followed-up for a minimum of six months.

No grade 3 or worse GU or GI toxicities were recorded. At 90 days post-treatment, cumulative acute Grade 2 GU and GI toxicities rates were 34% and 5%, respectively, making this focal boost regime, with an extreme hypofractionated basis, acceptable in terms of GU and GI toxicities.

The Hypo-FLAME trial was extended to examine the effects of bringing the OTT down from 29 to 15 days by treating bi-weekly (De Cock et al. 2023) (see Chapter 8, Table 8.1). Doses and SIB were maintained as for the previous study, and 124 intermediate-/high-risk patients were enrolled this time. Acute toxicity and QoL scores were compared with the original trial.

As with the previous trial, no Grade 3 or worse GU or GI toxicities were recorded. At 90 days post-treatment, cumulative acute Grade 2 GU and GI toxicity rates were approximately 48% and 7%, respectively. The GU toxicity was significantly greater in the bi-weekly than in the once-weekly treatments (34%); but GI toxicity was similar (5%). Those treated once weekly had superior acute bowel and urinary QoL scores. The bi-weekly schedule is therefore associated with acceptable acute GU and GI toxicity, but patients should be counselled on the short-term advantages of the longer (29-day) schedule. Data are needed for longer-term oncological outcomes.

In Chapter 8, Table 8.1, details are also noted of the trial conducted by Maas et al. (2023). This was a prospective trial with low-/intermediate-risk PCa patients reporting at 5 years (median follow-up 59.5 months) on toxicity, PROs on QoL and biochemical recurrence rates, following treatment with a five-fraction SBRT technique with SIB for a focal lesion. From the total of 26 patients enrolled, 23% were low-risk and 77% were intermediate-risk patients.

No biochemical failures were recorded. Approximately 38% experienced late Grade 2 GU and 12% late Grade 2 GI toxicities, whereas none experienced Grade 3 or higher toxicities. PRO QoL metrics were not significantly different from the pre-treatment baselines. Overall, the regimen showed excellent biochemical control and late GU/GI toxicities were not undue, while the QoL decrease was not judged to be significant.

For the MR-guided treatments, Leeman et al. (2023) have published a recent review and meta-analysis of acute toxicities comparing MR-guided adaptive treatments with FM- or CBCT-guided non-adaptive SBRT for PCa patients. For the review, 29 prospective studies were identified, with a sum total of over 2,500 patients. Inclusion criteria included patients on prospective clinical trials, evaluating SBRT for PCa, sample sizes greater than ten patients, SBRT delivered in four or five fractions with total doses of 35–45 Gy, photon-based treatments and documented acute toxicity profiles, published in or after 2018.

The studies examined showed a mean PTV margin of 4.2 mm (3–6 mm)/posteriorly 3.0 mm (3–3 mm) for the MR-guided studies. For the CT-guided studies, the margins were 4.5 mm (2–6 mm)/ posteriorly 3.3 mm (0–6 mm). Approximately 11% of the MR studies used a rectal spacer compared with 4% of the CT studies. Acute toxicity data was recorded typically up to 90 days or three months following SBRT (range 0–270 days), with no significant difference between the MR-guided and CT-guided SBRT studies. Pooled estimates for acute Grade 2 or higher GU and GI toxicities were 16% and 4%, respectively, for MR-guided adaptive treatments; for the CT-guided, non-adaptive treatments, the corresponding toxicities were 28% and 9%, respectively. Although longer follow-up analysis is needed to assess late toxicity and control outcomes, the MR-guided adaptive treatments already seem to suggest a significantly reduced risk of physician-assessed acute GU and GI toxicities, compared with CT-guided, non-adaptive SBRT treatments

Two studies of note for on-treatment MR-guided techniques are the MIRAGE study (Kishan et al. 2023), and the HERMES study (Westley et al. 2023). Both are described in detail in Chapter 8, Table 8.1.

The MIRAGE study is a non-blinded, single-centre, Phase 3 randomised trial comparing MR guidance (MR LINAC) with CT guidance for SBRT PCa patients treated with 40 Gy in five fractions (EOD). The report is based on 77 patients in the CT arm and 79 patients in the MR LINAC arm. The primary endpoint examined was acute toxicity (≤ 90 days after SBRT), with all patients having three months or more of follow-up.

Acute Grade 2 or greater GU toxicity was significantly lower in the MR arm compared with the CT arm (approximately 24% compared with 43%, respectively), with similar results for the acute Grade 2 or greater GI toxicity, with approximately 0% compared with 11%, respectively. EPIC-26 outcome measures also showed a significantly lower percentage of patients with clinically significant changes in bowel scores at one month, namely 25% for the MR-guided patients compared with 50% for the CT-guided patients. The initial findings showed a significant reduction in moderate acute physician-scored toxicity and a smaller decrement in patient-reported QoL in the MR-guided arm than the CT-guided arm (Kishan et al. 2023).

The HERMES trial (Westley et al. 2023) was a single-centre, noncomparative, randomised Phase 2 trial for intermediate- or (lower) high-risk PCa patients. Patients were allocated (1:1) to either a two-fraction (24 Gy) or five-fraction (36.25 Gy) treatments, with the two-fraction treatment involving a SIB to the MR-defined DIL. All patients (i.e., both groups) were treated on an MR LINAC with daily online ART.

The initial report was from the first ten patients in each group, reported 12 weeks post-treatment. Acute Grade 2 GU toxicity was reported in one patient (10%) in the five-fraction group and in two (20%) patients in the two-fraction group. No Grade 3 or worse toxicities have been reported so far, showing that both MR-guided adaptive regimes offer low levels of toxicity. Recruitment continues towards 23 participants in each group.

Finally, although MR-guided, another trial, 2SMART (Ong et al. 2023) (see Chapter 8, Table 8.1) is attempting to study a lower fractionation schedule. This is a Phase 2, single-arm, prospective trial examining the safety of two-fraction SABR treatments with a focal boost to the MR-defined DIL for low-/intermediate-risk PCa patients. A dose of 26 Gy (32 Gy to GTV focal boost) was delivered in two fractions, one week apart. Treatments are undertaken on a standard C-arm LINAC, using VMAT. The primary endpoint was minimal clinically important change (MCIC) in QoL within three months of the SABR treatment, assessed using EPIC-26 methodologies. Secondary endpoints were acute and late toxicities, PSA nadir and biochemical failure.

The study involved 30 men, 7% with low-risk disease and 93% with intermediate-risk PCa. With a median follow-up of 44 months (39–49 months), one patient (3%) had a biochemical failure, while MCIC assessment showed ten (33%), six (20%) and three (10%) patients with acute MCIC in urinary, bowel and sexual QoL domains, respectively. No Grade 3 or greater GU or GI toxicities were noted. The authors conclude that, overall, this two-fraction approach appears to be safe for dose escalation, with minimal impact on acute QoL.

REFERENCES

Alayed, Y., Davidson, M., Liu, S., et al. 2020. Evaluating the tolerability of a simultaneous focal boost to the gross tumor in prostate SABR: A toxicity and quality-of-life comparison of two prospective trials. *International Journal of Radiation Oncology, Biology, Physics*. 107: 136–142.

Aluwini, S., Pos, F., Schimmel, E., et al. 2015. Hypofractionated versus conventionally fractionated radiotherapy for patients with prostate cancer (HYPRO): Acute toxicity results from a randomised noninferiority phase 3 trial. *Lancet Oncology*. 16: 274–283.

Arcangeli, G., Saracino, B., Arcangeli, S., et al. 2017. Moderate hypofractionation in high-risk, organ confined prostate cancer: Final results of a phase III randomized trial. *Journal of Clinical Oncology*. 35: 1891–1897.

Catton, C., Lukka, H., Gu, C.-S., et al. 2007. Randomized trial of a hypofractionated radiation regimen for the treatment of localized prostate cancer. *Journal of Clinical Oncology*. 35: 1884–1890.

De Cock, L. Draulans, C., Pos, F., et al. 2023. From once-weekly to semi-weekly whole prostate gland stereotactic radiotherapy with focal boosting: Primary endpoint analysis of the multicentre phase II hypo-FLAME 2.0 trial. *Radiotherapy and Oncology*. 185: 109713.

Dearnaley, D., Syndikus, I., Mossop, H., et al. 2016. Conventional versus hypofractionated high-dose intensity-modulated radiotherapy for prostate cancer: 5-year outcomes of the randomised, non-inferiority, phase 3 CHHiP trial. *Lancet Oncology*. 17: 1047–1060.

Draulans, C., van der Heide, U., Haustermans, K., et al. 2020. Primary endpoint analysis of the multicentre phase II hypo-FLAME trial for intermediate and high risk prostate cancer. *Radiotherapy and Oncology*. 147: 92–98.

Fransson, P., Nilsson, P., Gunnlaugsson, A., et al. 2021. Ultra-hypofractionated versus conventionally fractionated radiotherapy for prostate cancer (HYPO-RTP-PC): Patient-reported quality-of-life outcomes of a randomised, controlled, non-inferiority, phase 3 trial. *Lancet Oncology*. 22: 235–245.

Ghadjar, P., Fiorino, C., af Rosenschold, P., et al. 2019. ESTRO ACROP consensus guideline on the use of image guided radiation therapy for localised prostate cancer. *Radiotherapy and Oncology*. 141: 5–13.

Hannan, R., Salamekh, S., Desai, N., et al. 2022. SABR for high-risk prostate cancer: A prospective multilevel MRI-based dose escalation trial. *International Journal of Radiation Oncology, Biology, Physics*. 113: 290–301.

Herrera, F., Valerio, M., Berthold, D., et al. 2019. 50-Gy stereotactic body radiation therapy to the dominant intraprostatic nodule: Results from a phase 1a/b trial. *International Journal of Radiation Oncology, Biology, Physics*. 103: 320–334.

Hoffman, K., Voong, K., Levy, L., et al. 2018. Randomized trial of hypofractionated, dose-escalated, intensity-modulated radiation therapy (IMRT) versus conventionally fractionated IMRT for localized prostate cancer. *Journal of Clinical Oncology*. 36: 2943–2949.

Jackson, W., Silva, J., Hartman, H., et al. 2019. Stereotactic body radiation therapy for localised prostate cancer: A systematic review and meta-analysis of over 6,000 patients treated on prospective studies. *International Journal of Radiation Oncology, Biology, Physics*. 104(4): 778–789.

Kishan, A., Dang, A., Katz, A., et al. 2019. Long-term outcomes of stereotactice body radiotherapy for low-risk and intermediate-risk prostate cancer. *Jama Network Open*. 2(2): e188006.

Kishan, A., Ma, T., Lamb, J., et al. 2023. Magnetic resonance imaging-guided vs computed tomography-guided stereotactic body radiotherapy for prostate cancer: The MIRAGE randomised clinical trial. *JAMA Oncology*. 9(3): 365–373.

Lee, W., Dignam, J., Amin, M., et al. 2016. Randomized phase III noninferiority study comparing two radiotherapy fractionation schedules in patients with low-risk prostate cancer. *Journal of Clinical Oncology*. 34: 2325–2332.

Leeman, J., Shin, K.-Y., Chen, Y.-H., et al. 2023. Acute toxicity comparison of magnetic resonance-guided adaptive versus fiducial or computed tomography-guided non-adaptive prostate stereotactic body radiotherapy: A systematic review and meta-analysis. *Cancer*. 129: 3044–3052.

Loblaw, A. 2020. Ultrahypofractionation should be a standard of care option for intermediate-risk prostate cancer. *Clinical Oncology*. 32: 170–174.

Lukka, H., Deshmukh, S., Bruner, D., et al. 2023. Fiver-year patient-reported outcomes in NRG oncology RTOG 0938, evaluating two ultrahypofractionated regimens for prostate cancer. *International Journal of Radiation Oncology, Biology, Physics*. 116(4): 770–778.

Lukka, H., Pugh, S., Bruner, D., et al. 2018. Patient reported outcomes in NRG oncology RTOG 0938, evaluating two ultrahypofractionated regimens for prostate cancer. *International Journal of Radiation Oncology, Biology, Physics*. 102(2): 287–295.

Maas, J., Dobelbower, M., Yang, E., et al. 2023. Prostate stereotactic body radiation therapy with a focal simultaneous integrated boost: 5-year toxicity and biochemical recurrence results from a prospective trial. *Practical Radiation Oncology*. 13: 466–474.

Marvaso, G., Gugliandolo, S., Bellerba, F., et al. 2020. Phase II prospective trial "Give Me Five" short-term high precision radiotherapy for early prostate cancer with simultaneous boost to the dominant intraprostatic lesion: The impact of toxicity on quality of life (AIRC IG-13218). *Medical Oncology*. 37: 74.

Murray, J., Tree, A., Alexander, E., et al. 2020. Standard and hypofractionated dose escalation to intraprostatic tumor nodules in localised prostate cancer: Efficacy and toxicity in the DELINEATE trial. *International Journal of Radiation Oncology, Biology, Physics*. 106(4): 715–724.

Ong, W., Cheung, P., Chung, H., et al. 2023. Two-fraction stereotactic ablative radiotherapy with simultaneous boost to MIR-defined dominant intra-prostatic lesion – Results from the 2SMART phase 2 trial. *Radiotherapy and Oncology*. 181: 109503.

Pollack, A., Walker, G., Horwitz, E., et al. 2013. Randomized trial of hypofractionated external-beam radiotherapy for prostate cancer. *Journal of Clinical Oncology*. 31: 3860–3868.

Tree, A., Ostler, P., van der Voet, H., et al. 2022. Intensity-modulated radiotherapy versus stereotactic body radiotherapy for prostate cancer (PACE-B): 2-year toxicity results from an open-label, randomised, phase 3, non-inferiority trial. *Lancet Oncology*. 23: 1308–1320.

Tree, A., Satchwell, L., Alexander, E., et al. 2023. Standard and hypofractionated dose escalation to intraprostatic tumor nodules in localised prostate cancer: 5-year efficacy and toxicity in the DELINEATE trial. *International Journal of Radiation Oncology, Biology, Physics*. 115(2): 305–316.

Van As, N., Tree, A., Patel, J., et al. 2023. 5-year outcomes from PACE-B: An international phase III randomised controlled trial comparing stereotactic body radiotherapy (SBRT) vs. conventionally fractionated or moderately hypo fractionated external beam radiotherapy for localised prostate cancer. *International Journal of Radiation Oncology, Biology, Physics*. 117(4): e2–e3.

Wang, S., Tang, W., Luo, H., et al. 2023. The role of image-guided radiotherapy in prostate cancer: A systematic review and meta-analysis. *Clinical and Translational Radiation Oncology*. 38: 81–89.

Westley, R., Biscombe, K., Dunlop, A. et al. 2024. Interim toxicity analysis from the randomised HERMES trial of 2- and 5-fraction magnetic resonance imaging-guided adaptive prostate radiation therapy. *International Journal of Radiation Oncology, Biology, Physics*. 118(3):682–687.

Widmark, A., Gunnlaugsson, A., Beckman, L., et al. 2019. Ultra-hypofractionated versus conventionally fractionated radiotherapy for prostate cancer: 5-year outcomes for the HYPO-RT-PC randomised, non-inferiority, phase 3 trial. *Lancet*. 394: 385–395.

Wolf, F., Sedlmayer, F., Aebersold, D., et al. 2021. Ultrahypofractionation of localised prostate cancer. *Strahlentherapie und Onkologie*. 197: 89–96.

Zapatero, A., Roch, M., Castro Tejero, P., et al. 2022. MRI-guided focal boost to dominant intraprostatic lesion using volumetric modulated arc therapy in prostate cancer. Results of a phase II trial. *British Journal of Radiology*. 95: 20210683.

Zhao, Y. Haworth, A., Rowshanfarzad, P., et al. 2023. Focal boost in prostate cancer radiotherapy: A review of planning studies and clinical trials. *Cancers*. 15: 4888.

10 Artificial Intelligence

10.1 INTRODUCTION

One cannot fail to notice the continuous developments happening in all manner of subject areas, in which artificial intelligence (AI) is a talking point. The development of it in science and medicine has been happening for years and has been introduced in some way into the radiotherapy pathway over many years now, mainly in terms of aiding reporting and diagnosis, but also within the direct radiotherapy pathway in terms of treatment planning. That development has continued into use in imaging for on-treatment image guidance and is at the heart (especially in terms of autosegmentation) of some of the latest developments in online adaptive radiotherapy (ART).

In this chapter, we will examine a sample of papers highlighting these particular areas of the influence of AI in radiotherapy as a whole, specifically for prostate cancer (PCa), and for image-guided radiotherapy (IGRT) and ART. Although much will be put forward describing where AI and machine learning (ML) methods likely have a place in radiotherapy, there are issues and challenges which must not be overlooked (Siddique and Chow 2020; Veeraraghavan and Deasy 2022), like, for instance, aspects of security in the acquisition, storage and access to big data; the needs to test and verify algorithms and where training needs to be robust with appropriate data and numbers of datasets; determination of the reasoning used in models which, at times, may be obscure; identification of bias in answers and highly uncertain answers, especially where there are cases quite different from those of training data; the lack of "common-sense" because of the lack of ground knowledge/rules/deductive processes; and the understanding of where responsibilities lie for when systems do not work as anticipated (Lahmi et al. 2022). References which the reader might like to start with are Siddique and Chow (2020) and Veeraraghavan and Deasy (2022) for AI in radiotherapy, Baydoun et al. (2023) and George et al. (2022) for AI in PCa, and Niu et al. (2022) for AI in image guidance.

10.2 AI AND RADIOTHERAPY

There are many parts of the cancer diagnosis and radiotherapy pathways where AI could have an influence, perhaps making processes quicker and more efficient and possibly improving upon expert judgements currently applied in large datasets, for which computer algorithms arguably are best placed to operate. There is considerable general potential for improving efficiency and consistency, perhaps beyond that of current clinical expertise (Veeraraghavan and Deasy 2022).

In the earliest part of the process of diagnosis, AI is at the heart of developments in computer-aided detection and diagnosis of cancer (Siddique and Chow 2020), with algorithms helping human experts through machine learning. Following imaging for radiotherapy planning, AI is aiding with image segmentation for normal tissues for treatment planning (on both computed tomography (CT) and magnetic resonance (MR) images) and also for on-treatment imaging (Veeraraghavan and Deasy 2022). Data-driven approaches are applied throughout the RT process, all of which could improve both quality and efficiency, such as to possibly aid outcome prediction; at CT and MR simulation; to help with clinical decision-making; to introduce ML and knowledge-based planning; applications to help with ART and plan validation; and to also work in areas like machine and process QA (Pillai et al. 2019). AI could aid in the quality assurance (QA) of processes in both supervised and unsupervised ways, to replace manual QA review, introduce guided methodology for QA of prostate treatment plans, using image, dose and region of interest (ROI) features for prediction, and change the process away from detecting errors and more towards decision guides if plans need

DOI: 10.1201/9781003050988-10

more or less attention in the QA process. These interventions could help optimise workload and improve the quality of pre-treatment checks (Pillai et al. 2019).

10.3 AI AND PCA IN GENERAL

AI has the potential to help optimise and improve processes at nearly every step in the cancer diagnosis and treatment pathway for PCa patients. Many models/studies still remain pre-clinical, needing a fuller and more robust verification and validation process, but, in recent times, robust models of great potential have emerged, many demonstrating improvements in efficiency and consistency which one wishes to see within healthcare in general and PCa treatments in particular (Baydoun et al. 2023). AI and deep learning (DL) methods bring the potential for validation across thousands of patients, given the nature and connectivity of big data sets worldwide within cancer and radiotherapy. To that end, work is still needed to design/implement new automated radiotherapy processes across multiple institutions and establish multi-disciplinary consortia for these tasks (Baydoun et al. 2023).

At the time of diagnosis, AI has shown promise in the pathology of biopsy (analysing results, in microscopic image analysis and consequently in risk stratification). In terms of core needle biopsies, the AI system neural networks, trained with core samples from over 1000 cases, have compared well with reference standards from multiple board-certified pathologists (George et al. 2022), and in radiological imaging (both aiding fusion between modalities and in diagnosis itself) with the consequence of helping to stratify risk and develop predictive biomarkers (Siddique and Chow 2020; George 2022; Baydoun et al. 2023).

Recent studies show promise and credible results for AI systems becoming an effective adjunct to pathologists' workflow for evaluating prostate tissue specimens (George et al. 2022). Some results have revealed comparable and similar results to those from expert pathologists for differentiating between malignant and benign cases, with some work from whole-slide imaging revealing high sensitivity and specificity (> 90% in each) in distinguishing between benign and malignant glands (George et al. 2022). Thus, there are AI applications for improving accuracy and efficiency in histopathology assessment and diagnostic image interpretation, risk-stratification (prognostication) and prediction of therapeutic benefit, potentially making treatments more personalised (George et al. 2022; Baydoun et al. 2023).

In terms of the radiotherapy-specific workflow for PCa patients, Baydoun and colleagues (2023) note some key areas for AI influence within the treatment planning process:

- Normally a minimum of seven days is needed between CT simulation and plan generation; transformational changes could be made through AI for efficiency of workflow by acceleration, automation, decrease in inter- and intrauser variability and operation refinement.
- For new MR-only workflows at pre-treatment, synthetic CT scans are needed for both dose calculations and for on-treatment reference image matching (synthetic CT scans and synthetic digitally reconstructed radiographs (DRRs) (see also Chapter 5). AI can help developments move on from bulk-density, probabilistic, voxel-to-voxel and atlas-based methods. DL methods recently validated seem to outperform previous methods (Baydoun et al. 2023).
- Within the planning process itself, there are now a number of academically and commercially AI-based software packages available for autocontouring/autosegmentation. Studies using retrospective data in simulated environments, evaluated using quantitative indices like Dice Similarity Coefficients (DSCs), Hausdorff distances and mean centre-of-mass distance, are showing great promise. More recent models and studies show the potential for automatic segmentation within organs themselves (i.e., beyond whole organs at risk (OARs), etc.) for transitional and peripheral zones within the prostate and for seminal vesicles (SVs) for diagnosis.

AI applications in PCa generally, and for RT in particular, remain highly promising; data exchange barriers remain but undoubtedly the technological and process advancements in validating thousands of pathology slides, prospective data acquisition and analysis from RCTs, and prospective use of new workflows for radical treatment are very positive. Work needs to continue to accrue robust AI-based evidence and results from AI-integrated clinical trials (Baydoun et al. 2023).

10.4 AI AND PCA IN DIAGNOSIS

In terms of diagnosis of PCa, there have been developments through clinical validation and implementation studies for AI aiding the diagnosis of prostate core needle biopsies, demonstrating that AI techniques can be effectively used for the automatization of needle biopsy screening, improving standardisation and quality control (QC) of cases. There have also been quality improvements in cancer grading using AI, although some findings suggest that data from multi-expert sources, rather than just 'patch-based' image training, are needed, as the latter was found to over-evaluate grading significantly when compared with a validation set which was patient-based (George et al. 2022). This emphasises some of the limitations with AI systems and the need for independent and appropriate verification with validation sets of data.

In terms of imaging, in particular MR, AI has been successfully adopted and incorporated alongside routine radiology practice, being able to aid and improve image registration and feature segmentation (e.g., from MR and ultrasound (US) images for estimating prostate gland volume) and for MRg biopsies, using AI software assistance. Different machine learning-based models, based on texture feature extraction from MR-derived apparent diffusion coefficient (ADC) maps, together with prostate specific antigen (PSA) biomarker information, are able to predict the chance of upgrade of Gleason grade given on biopsy post-radical prostatectomy with fairly accurate predictive proficiency (George et al. 2022).

For MR imaging, Turkbey and Haider (2022) and Wang et al. (2022) highlight areas where AI has improved MR scan quality, consistency, segmentation of structures (e.g., the prostate gland itself) and intraprostatic lesion detection.

For example, there are often large variations in MR scan quality, because of equipment used, pulse sequence parameters and patient-related factors (such as the presence of prostheses, internal organ motion, rectal gas, etc.). The Prostate Imaging-Reporting and Data System (PI-RADS) (Turkbey and Haider 2022; Wang et al. 2022) was developed in 2012 to improve reporting consistency and quality, although it does not necessarily ensure the quality of the MR scans themselves. Patient preparation measures (e.g., bowel preparation) have a role to play there. But AI can help with image quality and reporting quality aspects (Turkbey and Haider 2022; Wang et al. 2022). The structured format of PI-RADS and the standardisation it brings has been shown to improve sensitivity of multiparametric MR imaging (mpMRI) for clinically significant PCa. But, because even within PI-RADs, the core criteria are subjective for radiologists, applying AI can help support MR scan interpretation, particularly for mpMRI, a modality increasingly used with PCa patients, especially for feature segmentation in identifying and defining focal lesions and aiding in TNM staging, etc. (Turkbey and Haider 2022; Wang et al. 2022).

However, one must not underestimate the current limitations. AI systems require enormous amounts of training data sets or, if capable of self-training, require additional and expensive technological infrastructure support. Training needs to be performed with large-scale and diverse datasets, tested in real-life settings, before a full implementation into clinical practice (Turkbey and Haider 2022; George et al. 2022).

In terms of diagnosis alone, there is much to look forward to in the future. For example, AI use in fluorescence confocal microscopy and real-time optical imaging techniques processing fresh tissues and improvements in speed and efficiency of processing, when digital images are shared over the internet; this brings possibilities of real-time interpretation by remote pathologists as added benefits. The "one-hour-to-diagnosis" method is being evaluated on needle biopsies and results compared with conventional methods; AI methods could achieve approximately a 45-minute turnaround and > 90% sensitivity and specificity in diagnosis (George et al. 2022).

10.5 AI AND PCA IN TREATMENT PLANNING

The use of AI in treatment planning has already been mentioned. The use in autosegmentation/autocontouring alone could help in improving consistency and efficiency in plan generation.

McIntosh et al. (2021) led a study to investigate the integration of machine learning (ML) into the radical treatment of PCa patients. In a blinded, head-to-head study, integration into the clinical workflow was prospectively evaluated, comparing human and ML-generated plans directly. Both plans were physician-assessed in a blinded manner following *a priori* defined standardised criteria and peer review processes.

Overall, 89% of ML-generated plans were considered clinically acceptable, with 72% selected over human-generated RT plans in the head-to-head comparison. In terms of time savings, the ML approach reduced the median time required for the entire RT planning process by 60% (from 118 hours to 47 hours). However, acceptance for clinical deployment was not as positive, highlighting that the acceptance of the AI-based method in the real-world clinical setting, where patient-care is at stake, may still not be representative and requires greater study and investigation (McIntosh et al. 2021).

In a separate study, Hobbis et al. (2023) reported on the clinical acceptability of commercial deep learning- (DL-) based autosegmentation prostate models retrained using institutional data for clinical target volume (CTV) and OAR delineation from 109 post-prostatectomy patients. The performance of both the vendor-trained and the custom-retrained models was compared, with six clinical oncologists performing qualitative evaluations, scoring their preference and clinical utility for the structure sets. Consensus data were then used towards developing a separate CTV model. Hobbis et al. (2023) found that the custom retraining with the institutional data led to performance improvements, indicating that a small dataset could be sufficient for custom training the DL autosegmentation; for OARs from as little as 30 training datasets; for the CTV, as few as 60 datasets were needed for good performance. Although not without some examples of poor performance for both OARs and CTVs (see Figure 10.1), the results were considered good, with a small number of datasets being sufficient for institutional site-specific models. Further investigations continue, including evaluating possible reductions in interobserver variability.

Many studies demonstrate significant time savings (some studies indicate up to 50%, representing tens of minutes) in contouring compared with existing clinical methods, with acceptable geometric alignment (often comparing the automatic with the manually generated contours); dosimetric analysis also shows good agreement with manual approaches (McIntosh et al. 2021).

Through the advent of AI-based autosegmentation and autocontouring software, the National Institute for Health and Care Excellence (NICE) has conducted rapid formal appraisals of a number of published reports and studies of commercially available software packages for both OAR and target tissues (NICE 2023). From an assessment and evaluation of 79 reports (across all clinical sites, not just prostate) of eight full prospective papers, 19 full retrospective papers and 52 published conference abstracts, there was strong evidence for potential clinical usefulness for AI-based autocontouring, although unclear evidence of its cost-effectiveness. It was noted that studies were often poorly reported and sometimes lacking in clarity; sample sizes were small and most evidence was focussed predominantly on head and neck cancer and PCa. All studies reported either geometric, dosimetric or satisfaction scores which showed that AI-based autocontouring created contours, segmentations or plans similar to those created by manual methods for most OARs and target tissues. For some smaller OARs, autocontouring was more challenging suggesting its use as an aid to current clinical methods rather than as a stand-alone technology replacing those methods and roles (NICE 2023).

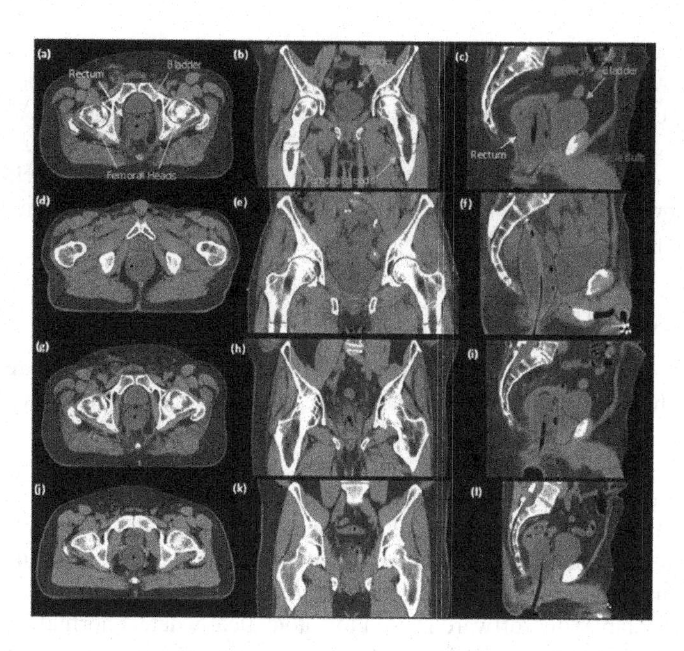

FIGURE 10.1 Examples of good and poor performance in AI-generated contours for OARs and CTVs. The sixty-case artificial intelligence model generated and the manual contours for bladder, rectum, femoral heads and penile bulb, showing one example of good performance (a, b, c) and one example of poor performance (d, e, f). The artificial intelligence 60-case model generated (and manual contours for the prostate bed clinical target volume, showing one example of good performance (g, h, i) and one example of poor performance (j, k, l) in three views. Image taken from Figure 4 of Hobbis, D., Yu, N., Mund, K., Duan, J., Rwigema, J-C., Wong, W., Schild, S., Keole, S., Feng, X., Chen, Q., Vargas, C., and Rong, Y. 2023. First report on physician assessment and clinical acceptability of custom-retrained artificial intelligence models for clinical target volume and organs-at-risk auto-delineation for postprostatectomy patients. *Practical Radiation Oncology*. 13: 351–362. Published by Elsevier Inc.

10.6 AI IN IMAGE GUIDANCE

AI has also been reported for its use in on-treatment image guidance, such as in contributing to cone-beam computed tomography (CBCT) image reconstruction in order to improve image quality, reduce noise contributions, reduce artefacts and lower effects of scatter contributions (Niu et al. 2022). Image registration and matching can be improved and made more efficient and consistent using DL approaches, building upon the success found in other imaging modalities in radiology (like CT, MR and PET) (Niu et al. 2022).

Synthetic CT has been mentioned before, produced for an MR-only workflow in pre-treatment planning, but similar AI applications can be used to generate synthetic CT from daily CBCT images as a basis for providing quantitative CT images with accurate Hounsfield unit (HU) values and anatomical information consistent with beam delivery time. Synthetic CT from CBCT images can be used to evaluate dose delivered (synthetic dose images) for the anatomy of the day. Consequently, anatomical information for contour delineation, dose calculation and target localisation can be performed as the heart of CBCT-based adaptive RT planning (Niu et al. 2022).

The advent of MR imaging at the point of treatment delivery, through MR LINACs, or indeed from MR-only workflows pre-treatment followed by X-ray-based IGRT on-treatment (on C-arm or O-ring LINACs) (see Chapter 5), brings forward the need for better image registration methods based upon soft-tissue structures. AI-based methods are now making these challenges a reality, as well as in terms of developing deformable models and registration to compute dose delivered throughout the treatment course (Teuwen et al. 2022). Some of the most time-consuming steps in current online ART processes (see Chapter 5), such as creation of synthetic CT from CBCT (as noted earlier), contour propagation and autosegmentation for target and OAR tissues, can be aided and made more efficient through AI-assisted methods (Teuwen et al. 2022). AI is likely to be the only way of creating fast enough algorithms for motion detection and tracking on-treatment (Siddique and Chow 2020; Teuwen et al. 2022), from real-time imaging and motion data for the ultimate of real-time ART.

10.7 AI AND ADAPTIVE RT

AI is very much at the heart of the online ART developments. The ones which are X-ray based, on O-ring LINAC technology (Ethos; see Chapter 5) are based on work by Archambault et al. (2020). This is work which is worth exploring in detail.

- The on-couch adaptive workflow developed is as follows:
  - kV CBCT acquisition, using iterative reconstruction methods (Kirby and Calder 2019), is acquired and presented for clinician evaluation as acceptable or requiring re-acquisition. This is the first decision step in the process.
  - Once acceptable, the software is used to automatically detect normal organ structures directly within the kV CBCT scan. Termed "influencer" structures, these are the ones closest in proximity to the target volume, therefore having the greatest impact on target shape and position. "Influencer" structures are presented to the clinician for review and adjustment where necessary until deemed adequate. This is the second decision step in the process.
  - "Influencer" structures are used to guide an algorithm that transfers target structures from the original plan or reference image into this accepted kV CBCT image set, thus ensuring that features and the relationships between the target and other anatomical structures originally present in the reference image are preserved within the kV-CBCT image space. Together, these propagated target and normal tissue structures constitute a new patient model, deemed the "Session model."' Reviewing and accepting this model is the third decision step.
  - The "Session model" is then used for the automatic generation of two plans: the first, using the original, scheduled (reference) plan applied to the "Session model" structures and anatomy; the second, a full, new optimisation and calculation derived from the session patient model. The two plans are presented and the user may choose one with which to continue treatment, namely the scheduled (reference) or the adapted plan. This is the fourth decision stage.
- Training, as mentioned earlier, is a vital part of AI and ML models, and here it involves three datasets:
  - **Training dataset** – used to fit the model. This is a large set of consistently contoured data used to perform the training of the neural network. Contours in training datasets are randomly peer-reviewed to ensure adherence to selected clinical guidelines.
  - **Validation dataset** – used to provide an unbiased evaluation of the model fitted from the training dataset while tuning the model further. The validation dataset is considered a subset of the training dataset.
  - **Test dataset** – a smaller set used to provide an evaluation of the final model fit. Scans related to a patient belonging to the test dataset CANNOT be used for network training. Each image and contour in the test dataset is reviewed by clinicians for accuracy and appropriateness.
- **Model verification and validation** – each trained neural network undergoes verification tests which are used to compare the obtained classification with ground truth contours on

multiple test datasets. Several evaluation metrics are often used, computed and evaluated during verification against passing criteria established for each of the evaluated structures. Only models passing verification are candidates for validation.

Models which pass verification are then validated by clinicians (clinical experts), in tests that better estimate the clinical review effort. Passing validation testing ensures that the model meets user needs and usability, which qualifies the model for clinical deployment.

This is the model and process which is part of Ethos, which had its CE marking and first patient treatment in 2019 in Herlev, Denmark. As a result, the Ethos therapy system brings in technology using AI and ML to create contours and generate adapted plans for clinicians to review, while a patient is on the treatment couch. Simple tools are offered within the software and at each decision-making step. Daily variations in the patient's anatomy, captured and visualised through iterative CBCT, enable an on-couch, on-treatment adaptive workflow, with initial results showing a process which enables an adapted plan to be created, chosen, quality assured and delivered in as little as 15 minutes. Papers giving independent evaluation and results are discussed in Chapter 5.

The other major technology for online ART is MR-guided ART (MRgART). Nachbar et al. (2023) have recently investigated AI-based autocontouring methods for PCa patients for MRgART through a study examining the training and validation of an AI-based model for autosegmentation on MR images as part of an MR LINAC/MRgART workflow. A total of 47 patients were included, each having T2w images acquired on the Elekta Unity MR LINAC on five different days. Structures that were manually outlined were prostate, SVs, rectum, anal canal, bladder, penile bulb, body and bony structures. These constituted the training data, amounting to 232 datasets in total, for generating a deep-learning autocontouring model. The model was then validated on 20 unseen T2w MRIs.

Quantitative evaluation was through the radiation oncologist contouring on the validation set, acting as the gold standard. MATLAB was used to compare the gold standard with automatic contours (AIC) using dice similarity coefficients (DSCs) and other quantitative metrics. Further qualitative evaluation was *via* five radiation oncologists independently scoring the AICs on possible usage within an online ART workflow, scoring them as

- (1) no modifications necessary
- (2) minor adjustments needed
- (3) major adjustments needed
- (4) unusable

For example, the results showed Max/Min DSC values of 97% for bladder and 73% for penile bulb. Qualitative evaluation showed a mean score of 1.2 for AICs across all organs and patients. For the different AICs, the highest mean score was 1.0 for the anal canal, sacrum, femur left and right, pelvis left; for the prostate, the lowest mean score was 2.0. A total of 80% of contours were rated as clinically acceptable, 16% requiring minor and 4% major adjustments for use in an online Adaptive MRgRT workflow. The median time for the AICs was 152 seconds (121–198 seconds).

Potential pitfalls were identified as the effect of images from different types of MR sequences; analysis here was only on fast MRIs (probably with less contrast between structures than some other sequences, a likely increased anatomical uncertainty). But, overall, it was concluded that the AI-based autocontouring was successfully trained for the online MRgART workflow, automatically generating contours which were found to be clinically acceptable in about 80% of cases or only needing minor modifications before clinical treatment.

10.8 AI AND ETHICS IN RADIOTHERAPY

Lahmi et al. (2022) present an interesting treatise on considering the ethics of AI within radiotherapy. In applying AI to so many parts of the Rt process, one might need to consider effects, such as those on human work –the potential challenges and dangers of de-skilling workers and the possibilities of automation bias. The consideration of optimum performance (for ourselves as human beings) is normally in a zone which balances out working within our comfort zone and aspects which enable growth and learning without raising stress levels too highly and moving individuals into burnout and anxiety.

Other issues include ourselves as experts and our expertise levels, and the maintenance of such. In addition, there are aspects of data quality and interpretability in terms of the AI models and understanding/being aware of the powers and limitations, and identifying the latter. Issues of liability in cases of error need careful and critical discussion and perhaps continuing development while respecting the "four topics" approach to medical ethics, ensuring AI processes operate in line with ethical values and their four pillars: medical indications, patient preferences, quality of life and contextual features.

At the same time, although one must not block AI for fear of being replaced, rather we should look to participate in its implementation to understand its issues and adapt to its presence. In the same way that AI needs training, so do we in understanding its potential and possible biases and the limitations of AI. Delegating does not mean total trust without control.

REFERENCES

Archambault, Y., Boylan, C., Bullock, D., et al. 2020. Making on-line adaptive radiotherapy possible using artificial intelligence and machine learning for efficient daily re-planning. *Medical Physics International Journal.* 8(2): 77–86.

Baydoun, A., Jia, A., Zaorsky, N., et al. 2023. Artificial intelligence applications in prostate cancer. *Prostate Cancer and Prostatic Diseases.* https://doi.org/10.1038/s41391-023-00684-0

George, R., Htoo, A., Cheng, M., et al. 2022. Artificial intelligence in prostate cancer: Definitions, current research and future directions. *Urologic Oncology: Seminars and Original Investigations.* 40: 262–270.

Hobbis, D., Yu, N., Mund, K., et al. 2023. First report on physician assessment and clinical acceptability of custom-retrained artificial intelligence models for clinical target volume and organs-at-risk auto-delineation for postprostatectomy patients. *Practical Radiation Oncology.* 13: 351–362.

Kirby, M., Calder, K-A. 2019. *On-treatment verification imaging: a study guide for IGRT.* Boca Raton, FL: CRC Press, Taylor and Francis Group.

Lahmi, L. Mamzer, M.-F., Burgun, A., et al. 2022. Ethical aspects of artificial intelligence in radiation oncology. *Seminars in Radiation Oncology.* 32: 442–448.

McIntosh, C., Conroy, L., Tjong, M. 2021. Clinical integration of machine learning for curative-intent radiation treatment of patients with prostate cancer. *Nature Medicine.* 27: 999–1005.

Nachbar, M., Lo Russo, M., Gani, C. et al. 2023. Automatic AI-based contouring of prostate MRI for online adaptive radiotherapy. *Zeitschrift für Medizinische Physik.* S0939-3889(23)00053-3.

National Institute for CE (NICE). 2023. *Artificial intelligence auto-contouring to aid radiotherapy treatment planning [GID-HTE10015].* External Assessment Group report. Exeter: Peninsula Technology Assessment Group (PenTAG), University of Exeter Medical School.

Niu, T., Tsui, T., Zhao, W. 2022. AI-augmented images for X-ray guiding radiation therapy delivery. *Seminars in Radiation Oncology.* 32: 365–376.

Pillai, M., Adapa, K., Das, S., et al. 2019. Using artificial intelligence to improve the quality and safety of radiation therapy. *Journal of the American College of Radiology.* 16: 1267–1272.

Siddique, S., Chow, J. 2020. Artificial intelligence in radiotherapy. *Reports of Practical Oncology and Radiotherapy.* 25: 656–666.

Teuwen, J., Gouw, Z., Sonke, J.-J. 2022. Artificial intelligence for image registration in radiation oncology. *Seminars in Radiation Oncology.* 32: 330–342.

Turkbey, B., Haider, M. 2022. Deep learning-based artificial intelligence applications in prostate MRI: Brief summary. *British Journal of Radiology.* 95: 20210563.

Veeraraghavan, H., Deasy, J. 2022. Chapter 11: Artificial intelligence in radiation oncology: A rapidly evolving picture. In *Image-guided high-precision radiotherapy.* Ed. E. Troost. Cham: Springer Nature, 249–270.

Wang, L., Margoglis, D., Chen, M. 2022. Quality in MR reporting of the prostate – Improving acquisition, the role of AI and future perspectives. *British Journal of Radiology.* https://doi.org/10.1259/bjr.20210816

11 Training

11.1 INTRODUCTION

A key aspect of any clinical insight into Image-guided Radiotherapy (IGRT) – for all sites, not just prostate – is the aspect of training and competency. From the original national guidance for IGRT (RCR 2008; NRIG 2012), this was stressed, from the earliest of on-treatment imaging equipment using only MV portal imaging and the first generation of Electronic Portal Imaging Devices (EPIDs), through to the advent of amorphous silicon flat panel imagers and subsequent introduction of kV Cone-beam Computed Tomography (kV-CBCT) on C-arm LINACs, and also for all integrated image guidance on other advanced radiotherapy technologies (Kirby and Calder 2019, chapters 4 and 12).

The latest national guidance (RCR 2021) continues to emphasise this need for training and education, with well-designed and structured competencies within the clinical department. Interprofessional IGRT training is highlighted because of the multidisciplinary nature of modern IGRT. In the UK, on-treatment IGRT is therapeutic radiographer-led as part of a wider staff team, as is emphasised with the newer practices in online adaptive radiotherapy (ART) (see Chapter 5).

Training should be discipline-specific to best suit the different roles and responsibilities, which may vary in different departments and staffing models. It is something that must evolve as technology changes and practice, service requirements and experiences develop. Ideally, there should be IGRT specialists within departments, within the different professions involved. With all treatment equipment having some form of IGRT for on-treatment imaging, all therapeutic radiographers and other staff require ongoing training and appropriate skills and competencies. Typical training requirements and content are outlined fully in On-Target 2 (RCR 2021, chapter 9). General training in IGRT should be alongside site-specific training, which is one of the reasons for the existence of this book!

Training methods can vary, depending on resources, equipment, techniques and levels of training required. Some of the newer methods mentioned below are becoming part of regular clinical training and competency; certainly, innovative pedagogy is being used through simulation in the pre-registration training environment (for all disciplines), but also with registered professionals and importantly for patient education. Patient compliance is a key issue for accurate and precise radiotherapy to be effective and this can only come through greater patient knowledge and education for all aspects of their treatment. A variety of formats are mentioned for consideration in On-Target 2 (RCR 2021), together with methods for assessing competency and maintaining/developing it further throughout a professional career. The rationale of training, together with suggested methods and resources, are available in chapters 4 and 12 in Kirby and Calder (2019) and chapter 9 of On-Target 2 (RCR 2021). The reader is encouraged to consult these sections first if they are unfamiliar with them.

11.2 PRE-REGISTRATION TRAINING

The ideal training for IGRT, for all radiotherapy training and certainly for site-specific oncology and radiotherapy for prostate cancer (PCa) patients begins with pre-registration training for all cancer professionals – therapeutic radiographers, radiotherapy physicists, clinical oncologists, dosimetrists, oncology nurses and radiotherapy engineers and technicians.

The methods used have evolved from just didactic classroom experiences and self-directed study to a fully blended learning and teaching approach, especially in the healthcare sciences. Within our own experience, this now involves a full range of pedagogic techniques with the stress on simulation

DOI: 10.1201/9781003050988-11

and interprofessional learning, bringing practical, visual, engaging and kinaesthetic approaches to learning. Most educational institutions combine e-learning, multimedia, virtual reality and online sources with traditional lectures, tutorials, group work, seminars and one-to-one supervision, to name but a few. These methods are used (for example) for teaching the science behind radiotherapy and IGRT (Kirby 2018; Jimenez et al. 2018a; Bridge et al. 2020); helping students to understand aspects of treatment planning, the implications of high dose-gradients and conformality, margins and therefore the need for IGRT.

Simulation is now an integral part of therapeutic radiography pre-registration teaching programmes in the UK (Ketterer et al. 2020), and also nationally and internationally for the medical radiation sciences (Bridge et al. 2021), bringing multiple disciplines together to learn about radiotherapy and the interdisciplinary connections needed for the modern effective healthcare team (Jimenez et al. 2018a; Ball et al. 2021), including wider professions in healthcare (Bridge et al. 2022) – especially where the use of simulation can alleviate some of the pressures and burdens found in teaching in busy clinical environments and still be effective.

Simulation has developed to such an extent that there is now national guidance on its continuing development and use within radiotherapy, but also in healthcare as a whole (SOR 2022). From the national study and published guidance for pre-registration education for therapeutic radiographers, in particular, one of the four overarching themes where education should focus simulation efforts to ensure the optimum and the greatest benefit for student experience and learning was "Treatment and Imaging", including the main aspects within the HCPC standards of proficiency (HCPC 2023) for the full range of radiotherapy processes and techniques for safe and accurate patient set-up. This is at the heart of IGRT training, together with developing 2D and 3D interpretation and evaluation skills for accurate dose delivery (SOR 2022).

In addition, within our own pre-registration training, all aspects of treatment planning, verification and delivery are taught to our undergraduate and postgraduate students; using both fundamental scientific principles in real clinical scenarios and also a case-based approach to site-specific oncology and radiotherapy teaching. IGRT is fundamental to this teaching, technically and clinically, for photon and proton beam therapy and for the newer advanced and emerging technologies. Kirby and Calder (2019) was written as the study guide for IGRT (and all associated aspects) covered throughout our undergraduate and postgraduate pre-registration courses, being written as a textbook to accompany the teaching. Only a few site-specific examples are contained within it, which is why this book and others are being offered, for pre-registration students and also healthcare professionals in their first clinical posts.

11.3 USE OF NEWER TRAINING METHODS – VR AND ELEARNING

Many new methods of teaching involve those mentioned earlier, but perhaps, technically, the biggest innovation is the use of virtual reality (VR), specifically through commercially available software like the Virtual Environment for Radiotherapy Training (VERT). This is a type of eLearning environment which we (and other researchers across the world) have found helps students engage better in learning and teaching key radiotherapy, radiotherapy physics and practical clinical aspects, simulating the real clinical environment well and allowing students to learn interactively and dynamically in a safe and less-pressured way than the clinical workplace; but in a manner which has been proven to be highly effective. With full 3D immersive models (see Figure 11.1) of some of the most up-to-date commercially available radiotherapy equipment (for both photons and protons), using control pendants and screen displays actually used on clinical equipment, the system is an integral part of teaching within universities and the majority of cancer hospitals in the UK.

Some examples of papers outlining its use and how that has grown over the years include examining the international landscape of academic practice using VERT (Bridge et al. 2017); explaining its development and the opportunities of this type of simulation training in radiotherapy (Beavis 2018):

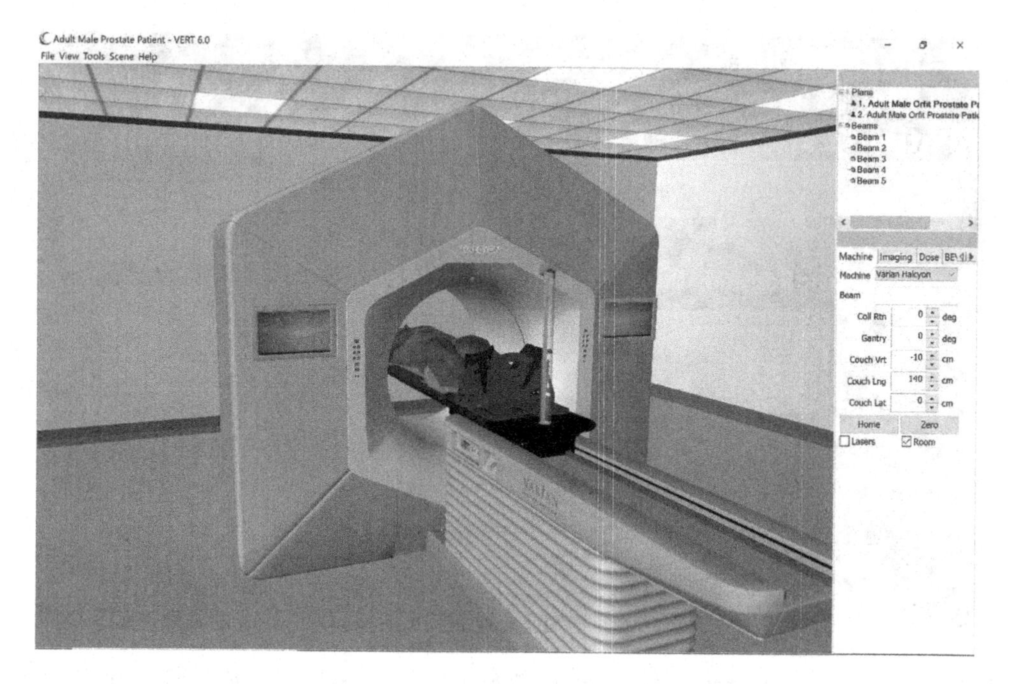

FIGURE 11.1 VERT environment for simulating new equipment: Varian's Halcyon. Image courtesy of Prof. Andy Beavis/Vertual Ltd.

reviewing its use in simulation-based education (Kane 2018); outlining its use for patient education in the clinic (Jimenez and Lewis 2018b, 2018c); illustrating how it can be used to support undergraduate teaching for modern radiotherapy techniques and quality assurance (QA) (Chamunyonga et al. 2018); and examining its use for medical dosimetrist training (Cheung et al. 2021); sharing experiences of use for training, including IGRT and CBCT (Wijeysingha et al. 2021); and evaluating new features such as treatment planning evaluation tools and how they can be used in teaching environments (Bridge et al. 2020; Ramashia 2023). In many cases, VERT and other e-learning and VR environments cover aspects pertinent to good radiotherapy practice in general and IGRT learning in particular.

Self-directed eLearning packages also have their place – some have been in design and use for many years as part of the NHS England eLearning for Health initiative (NHS England 2023a) and include IGRT within other advanced radiotherapy topics. Some can be bespoke with specific radiotherapy learning objectives tailored for the needs of a particular department/hospital trust (Oliver 2021; Kirby 2021) (see Figures 11.2 and 11.3), helping both therapeutic radiographers and other healthcare professionals support patients better through and beyond their treatment course. As research develops for the newer forms of IGRT and online ART, greater patient information on likely side effects of PCa treatment has been suggested (Teunissen et al. 2023); eLearning packages like those mentioned could very easily be designed for patient use during and following treatment, through multimedia and computer/tablet/smart phone applications.

Within the VERT software, IGRT tools have been specifically designed to simulate image acquisition and matching interfaces for different types of radiotherapy equipment and for prostate and other clinical sites (Figures 11.4 and 11.5). Evaluations have been published in recent times (Chamunyonga 2020, 2021; Stewart-Lord 2022), showing that it is a sound, suitable and realistic platform for training purposes, especially for image-matching skills. Not without its limitations, it provides opportunities for staff to increase their confidence in using matching software and technology, although evaluation skills need feedback and further supervision, provided through the educational programmes.

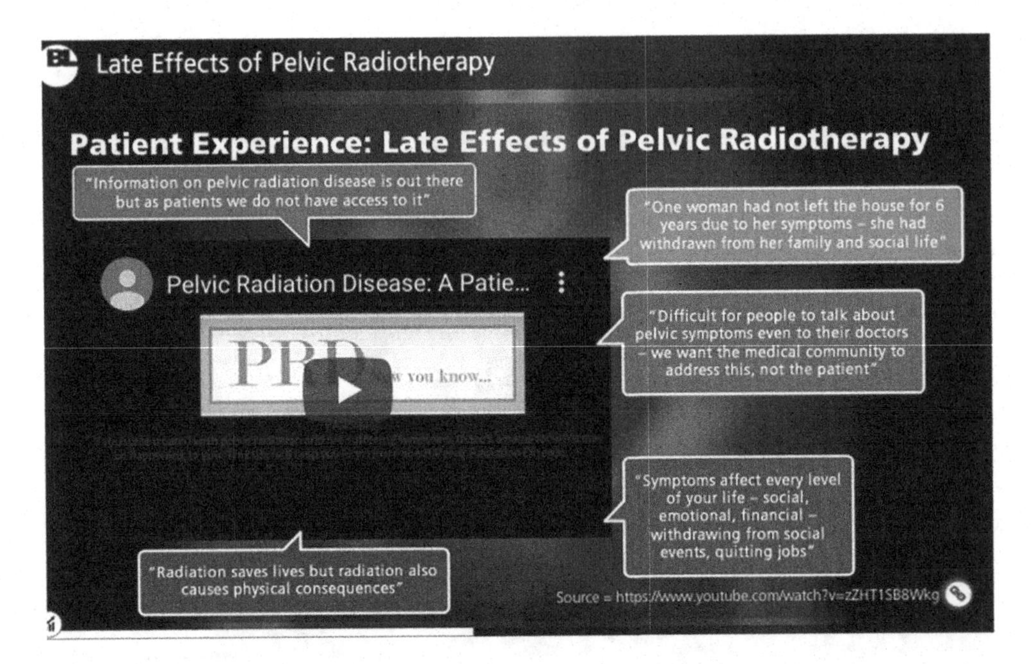

FIGURE 11.2 Example of an eLearning package. This one is used for helping professionals understand further the late effects from pelvic radiotherapy and is especially useful for PCa patients. Sample content contains links to patient experience video clips. Image taken from Figure 1 of Oliver L.A., Porritt B., Kirby M. A novel e-learning tool to improve knowledge and awareness of pelvic radiotherapy late effects: qualitative responses amongst therapeutic radiographers. *BJR Open* 2021; 3: 20210036.

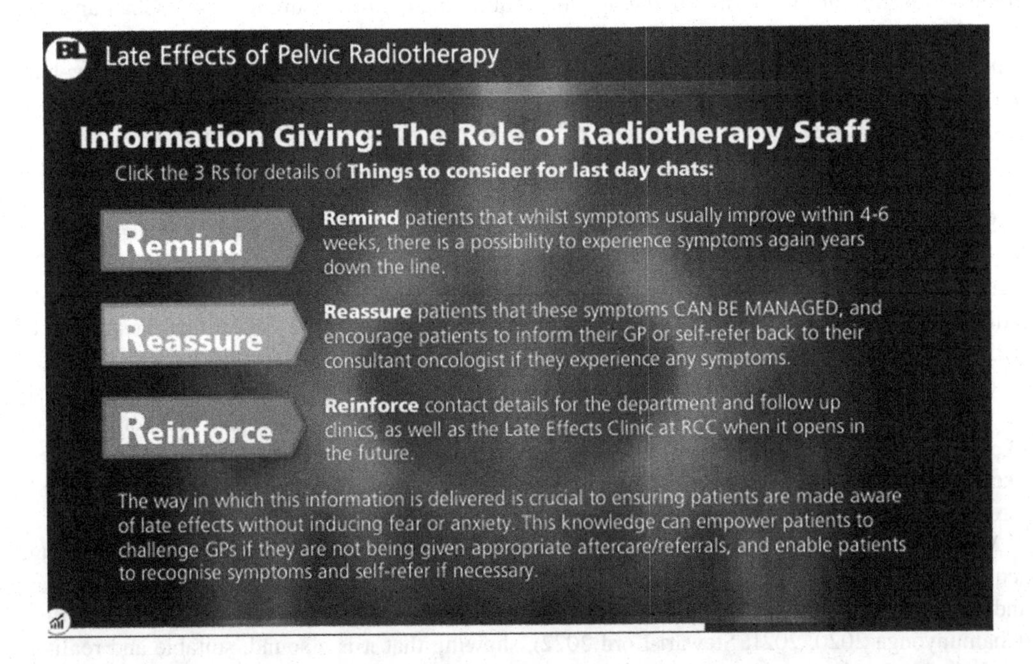

FIGURE 11.3 Example of an eLearning package. This one is used for helping professionals with tips and reminders for things to consider for pelvic radiotherapy patients (including PCa patients) on their last-day chats. Image taken from Figure 4 of Oliver L.A., Porritt B., Kirby M. A novel e-learning tool to improve knowledge and awareness of pelvic radiotherapy late effects: qualitative responses amongst therapeutic radiographers. *BJR Open* 2021; 3: 20210036.

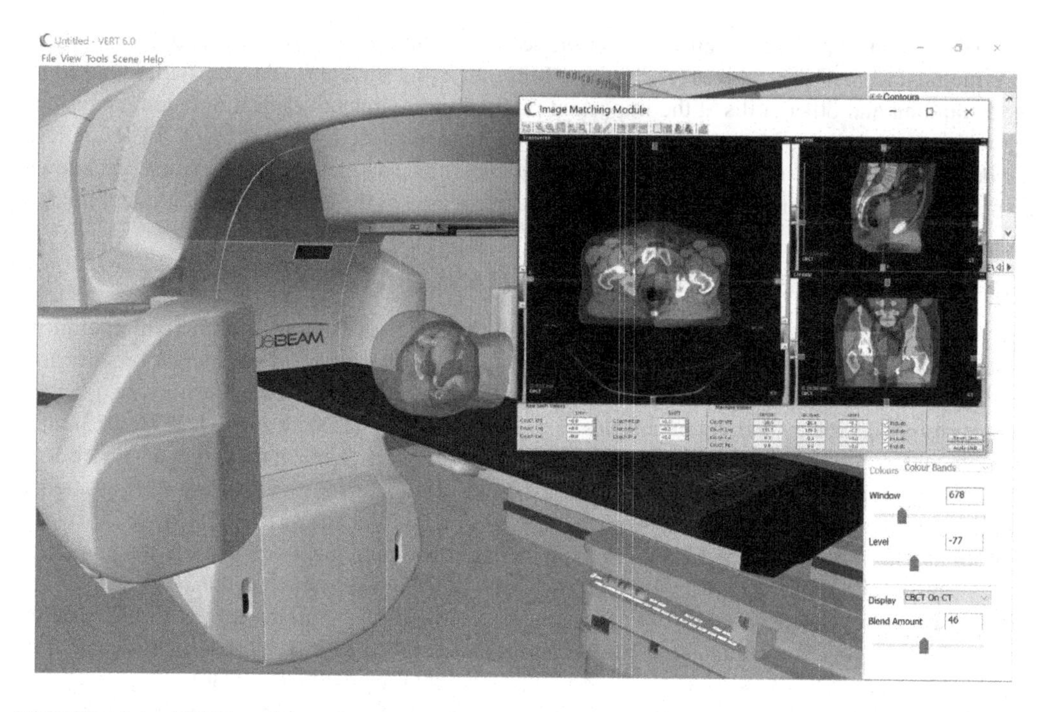

FIGURE 11.4 VERT environment for simulating and teaching prostate IGRT-CBCT acquisition, image matching, set-up correction and treatment delivery on Varian equipment. Image courtesy of Prof. Andy Beavis/Vertual Ltd.

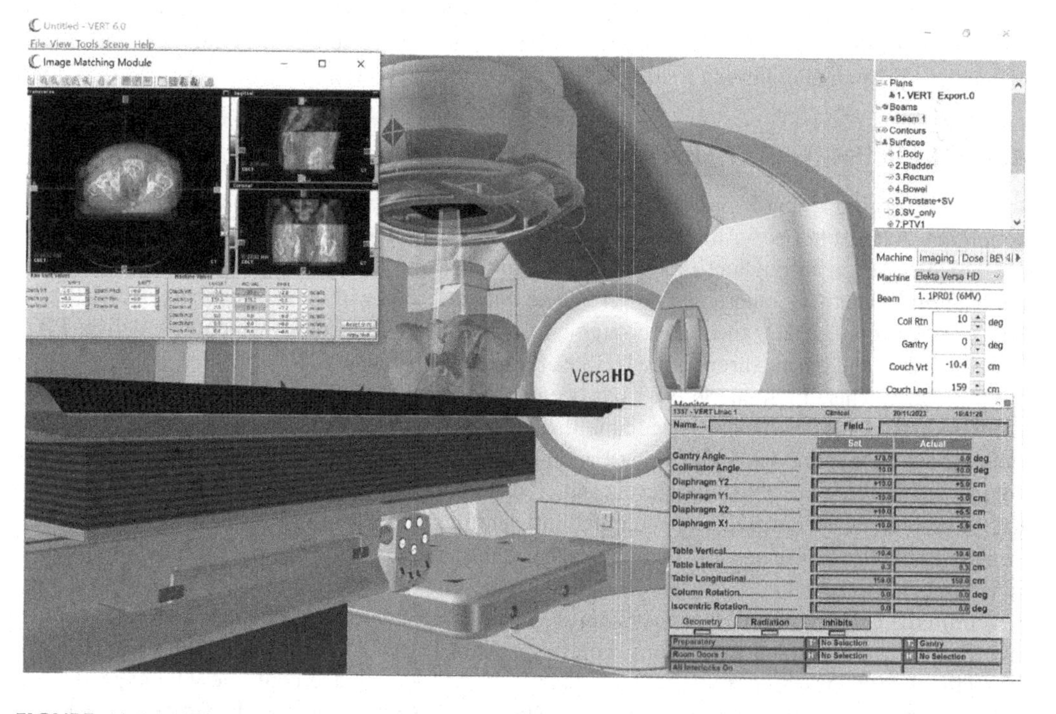

FIGURE 11.5 VERT environment for simulating and teaching prostate IGRT-CBCT acquisition, image matching, set-up correction and treatment delivery on Elekta equipment. Image courtesy of Prof. Andy Beavis/Vertual Ltd.

Further extensions to the simulation environments are continuing to be developed, one of which is a simulated LINAC control area (Kirby et al. 2022), allowing students to develop their knowledge, understanding and other skills at the important point of on-treatment verification and final checks before beam switch-on. Many different areas of simulation (including VERT and VERT IGRT) can be brought into use in simulated clinical scenarios, which can include set-up errors, breakdown events and other situations normally found in the control area of the LINAC in real clinical life.

11.4 EXPERIENCE FROM OTHER USERS: MANUFACTURER'S COURSES

Learning for radiotherapy and IGRT also come from the experiences of other users in their clinical implementation, sometimes when they are the first to have a new piece of equipment or software. Ideally shared as peer-reviewed research and development articles, valuable learning also comes from web-based articles and commentaries (e.g., SOR 2021; RFH 2023; NHS RSCH 2023), and through conference proceedings and posters (e.g., Bolt et al. 2021). Some of the larger and more academically experienced departments have well-established courses for radiotherapy staff to attend, usually run each year and employing world experts within their faculties (ICR 2023a, 2023b).

With any new piece of LINAC, radiotherapy and/or IGRT equipment or software, the manufacturer's courses and training have a vital role to play (RCR 2021; Kirby and Calder 2019, chapter 13) within the commissioning process, often enabling cascade training to many staff. Online resources help greatly in providing ready access to such materials, as well as in-person and online training opportunities run by the vendors (e.g., Varian 2023a, 2023b; Elekta 2023a, 2023b).

Formal training and development is also offered to healthcare professionals in service through university-based continuous professional development (CPD) courses and postgraduate qualifications (e.g., UoL 2023). CPD is a requirement of the Health and Care Professions Council (HCPC) registration in the UK (HCPC 2017) for many healthcare professions associated with radiotherapy and many of the disciplines needed for the provision of IGRT. CPD is the main method by which those in UK healthcare professions continue to learn and develop so that they can keep skills and knowledge updated, maintaining a safe and effective clinical service.

11.5 TRAINING FOR OTHER MODALITIES

New and emerging technologies are always on the horizon, including the use of other treatment modalities. A notable one in the UK is that of high-energy proton beam therapy (PBT), a subject which is taught throughout therapeutic radiographer and radiotherapy physicist pre-registration training, sometimes embracing newer educational methods within the university, like VR and VERT (Rabus et al. 2021), and also through clinical centres and faculties which now deliver PBT (Christie 2023) and online through the NHS eLfH hub (NHS England 2023b).

In the USA, where high-energy PBT has been in clinical use for decades, there is a consideration that only now is PBT sufficiently available that training and learning in radiation oncology must encompass PBT as a modality and at least an alternative modality (Wallner et al. 2023). Increasing clinical rotations are possible (at all levels and all disciplines) through PBT facilities; one might say that that should be encouraged and integrated into learning and teaching programmes at all clinical training levels (for physicians, medical physicists and radiation therapists) and throughout professional careers.

Within the US programmes, suggestions are that, during board assessments, examinees should be expected to understand indications, contraindications, general techniques and seminal clinical trials of PBT (Wallner et al. 2023). A higher level of knowledge could now be expected for qualifying and certifying examinations in the US. As an example, in the UK, further knowledge of particle therapy is expected within pre-registration programmes as standards of proficiency (HCPC 2023); this is especially important, one might note, for facilities with PBT and/or a higher proportion of referrals for, e.g., paediatric solid tumours and CNS and base-of-skull lesions. Even those not able to be rotated through PBT facilities should still be expected to understand it as an alternative modality.

11.6 TRAINING AND WORKFORCE DEVELOPMENT

As we have seen in Chapter 5, the more advanced online ART techniques (both X-ray- and MR-based) are resource hungry in terms of personnel, training and development. The resources required include the availability at the treatment console of the multidisciplinary team, each with appropriate high-level skills and experience, for performing online adaptive tasks and approval of contours/plans before treatment delivery (Hales et al. 2020; McNair et al. 2020, 2021; RCR 2021, p. 74; Zwart et al. 2022; Byrne et al. 2022; McComas et al. 2023; Smith et al. 2023).

Extra training needs have been identified comprehensively for those centres implementing online ART (Hales et al. 2020; McNair et al. 2021), representing additional knowledge, learning and skills beyond those used for traditional and conventional IGRT. Although intensive and still relatively new, work is already underway to establish newer and possibly more efficient ways of working, to reduce the staff resources required and examine opportunities for developing and enhancing the roles and responsibilities of other staff, as proven through robust research to be able to provide equivalent safety, accuracy and precision to online ART as current multidisciplinary models (Smith et al. 2023).

REFERENCES

Ball, B., Kirby, M., Ketterer, S-J., et al. 2021. Radiotherapy-specific interprofessional learning through simulation. *Radiography*. 27: 187–192.

Beavis, A. 2018. The opportunities of computer simulation training in radiation therapy. *Journal of Medical Radiation Sciences*. 65: 77–79.

Bolt, M., Shelley, C., Hollingdale, R., et al. 2021. *Assessment of automated radiotherapy plan generation for bladder cancer using the Ethos TPS*. Presented at the BIR Annual Radiotherapy and Oncology Meeting Part 2, 24–26 March 2021, Online (Virtual event).

Bridge, P., Adeoye, J., Edge, C., et al. 2022. Simulated placements as partial replacement of clinical training time: A Delphi consensus study. *Clinical Simulation in Nursing*. 68: 42–48.

Bridge, P., Giles, E., Williams, A., et al. 2017. International audit of VERT academic practice. *Journal of Radiotherapy in Practice*. 16(4): 375–382.

Bridge, P., Kirby, M., Callender, J. 2020. Evaluating VERT as a radiotherapy plan evaluation tool: Comparison with treatment planning software. *Journal of Radiotherapy in Practice*. 19: 210–214.

Bridge, P., Shiner, N., Bolderston, A., et al. 2021. International audit of simulation use in pre-registration medical radiation science training. *Radiography*. 27: 1172–1178.

Byrne, M., Archibald-Heeren, B., Hu, Y., et al. 2022. Varian ethos online adaptive radiotherapy for prostate cancer: Early results of contouring accuracy, treatment plan quality, and treatment time. *Journal of Applied Clinical Medical Physics*. 23:e13479.

Chamunyonga, C., Burbery, J., Caldwell, P., et al. 2018. Utilising the virtual environment for radiotherapy training system to support undergraduate teaching of IMRT, VMAT, DCAT treatment planning, and QA concepts. *Journal of Medical Imaging and Radiation Sciences*. 49: 31–38.

Chamunyonga, C., Rutledge, P., Caldwell, P., et al. 2020. The application of the virtual environment for radiotherapy training to strengthen IGRT Education. *Journal of Medical Imaging and Radiation Sciences*. 51: 207–213.

Chamunyonga, C., Rutledge, P., Caldwell, P., et al. 2021. The implementation of MOSAIQ-based image-guided radiation therapy image matching within radiation therapy education. *Journal of Medical Radiation Sciences*. 68: 86–90.

Cheung, E., Law, M., Cheung, F. 2021. The role of virtual environment for radiotherapy training (VERT) in medical dosimetry education. *Journal of Cancer Education*. 36: 271–277.

Christie (The Christie NHS Foundation Trust). 2023. *The Christie proton school*. Available at https://www.christie.nhs.uk/education/departments/the-christie-proton-school (accessed on 28 November 2023).

Elekta. 2023a. *Elekta unity*. Available at https://www.elekta.com/products/radiation-therapy/unity/ (accessed on 28 November 2023).

Elekta. 2023b. *MR-Linac mastercourse: Elekta unity clinical experience*. Available at https://webinars.elekta.com/watch/aJg73D9JAk8mX39aBPnTs3 (accessed on 28 November 2023).

Hales, R., Rodgers, J. Whiteside, L., et al. 2020. Therapeutic radiographers at the helm: Moving towards radiographer-led MR-guided radiotherapy. *Journal of medical imaging and radiation sciences*. 51: 364–372.

HCPC (Health and Care Professions Council). 2017. *Continuing professional development and your registration*. London: The Health and Care Professions Council.

HCPC (Health and Care Professions Council). 2023. *Standards of proficiency – Radiographers*. London: The Health and Care Professions Council.

ICR (Institute of Cancer Research). 2023a. *Image guided and adaptive radiotherapy in clinical practice: Including SABR topics course*. Available at: https://www.icr.ac.uk/studying-and-training/opportunities-for-clinicians/radiotherapy-and-imaging-training-courses/image-guided-intensity-modulated-radiotherapy-in-clinical-practice-course (accessed on 28 November 2023).

ICR (Institute of Cancer Research). 2023b. *Magnetic Resonance Image Guided Radiotherapy (MRIgRT) course*. Available at: https://www.icr.ac.uk/studying-and-training/opportunities-for-clinicians/radiotherapy-and-imaging-training-courses/magnetic-resonance-image-guided-radiotherapy-(mrigrt) (accessed on 28 November 2023).

Jimenez, Y., Thwaites, D., Juneja, P., et al. 2018a. Interprofessional education: Evaluation of a radiation therapy and medical physics student simulation workshop. *Journal of Medical Radiation Sciences*. 65: 106–113.

Jimenez, Y., Lewis, S. 2018b. Radiation therapy patient education using VERT: Combination of technology with human care. *Journal of Medical Radiation Sciences*. 65: 158–162.

Jimenez, Y., Lewis, S. 2018c. Radiation therapy patient education review and a case study using the virtual environment for radiotherapy training system. *Journal of Medical Imaging and Radiation Sciences*. 49: 106–117.

Kane, P. 2018. Simulation-based education: A narrative review of the use of VERT in radiation therapy education. *Journal of Medical Radiation Sciences*. 65: 131–136.

Ketterer, S.-J., Callender, J. Warren, M., et al. 2020. Simulated versus traditional therapeutic radiography placements: A randomised controlled trial. *Radiography*. 26: 140–146.

Kirby, M. 2018. The VERT Physics environment for teaching radiotherapy physics concepts - Update of four years' experience. *Medical Physics International Journal*. 6(2): 247–54.

Kirby, M. 2019. *Simulation and VR in Education: Experiences with Radiotherapy/Radiotherapy Physics and 'food-for-thought' from other disciplines*. Presented at MPEC 2019 IPEM Medical Physics and Engineering Conference, 23–25 September 2019, Bristol.

Kirby, M., Calder, K-A. 2019. *On-treatment verification imaging: A study guide for IGRT*. Boca Raton, FL: CRC Press, Taylor and Francis Group.

Kirby, M., Oliver, L., Kind, L., et al. 2021. A novel eLearning tool for ongoing radiation therapist education on pelvic radiotherapy late effects. *Radiotherapy and Oncology*. 161(S1): S227–S228.

Kirby, M., Porritt, B., Calder, K-A., et al. 2022. The design and construction of a simulated linac control area (SLCA) for radiation therapy. *Radiotherapy and Oncology*. 170(S1): S589–S590.

McComas, K., Yock, A., Darrow, K., et al. 2023. Online adaptive radiation therapy and opportunity cost. *Advances in Radiation Oncology*. 8: 101034.

McNair, H. Joyce, E., O'Gara, G., et al. 2021. Radiographer-led online image guided adaptive radiotherapy: A qualitative investigation of the therapeutic radiographer role. *Radiography*. 27: 1085–1093.

McNair, H., Wiseman, T., Joyce, E., et al. 2020. International survey; current practice in On-line adaptive radiotherapy (ART) delivered using Magnetic Resonance Image (MRI) guidance. *Technical Innovations and Patient Support in Radiation Oncology*. 16: 1–9.

NHS England (National Health Service England). 2023a. *Radiotherap-e: e-learning for advanced radiotherapy techniques*. Available at: https://www.e-lfh.org.uk/programmes/advanced-radiotherapy/ (accessed on 28 November 2023).

NHS England (National Health Service England). 2023b. *Proton beam therapy (eProton)*. Available at: https://www.e-lfh.org.uk/programmes/proton-beam-therapy/ (accessed on 28 November 2023).

NHS RSCH (Royal Surrey County Hospital). 2023. *Adaptive radiotherapy using Varian ETHOS*. Available at: https://medphys.royalsurrey.nhs.uk/department/radiotherapy-physics/research-in-radiotherapy/adaptive-radiotherapy-using-varian-ethos/ (accessed on 28 November 2023).

NRIG (National Radiotherapy Implementation Group). 2012. *Image guided radiotherapy— guidance for implementation and use*. London: National Cancer Action Team. 42. Oliver, L., Porritt, B., Kirby, M. 2021. A novel e-learning tool to improve knowledge and awareness of pelvic radiotherapy late effects: qualitative responses amongst therapeutic radiographers. *BJR Open*. 3: 20210036.

Rabus, A., Kirby, M., Nasole, L., et al. 2021. Evaluation of a VERT-based module for proton radiotherapy education and training. *Journal of Radiotherapy in Practice*. 20: 139–143.

Ramashia, P. 2023. Radiotherapy plan evaluation tool in a resource-limited setting: Comparison of VERT and treatment planning software. *Journal of Medical Imaging and Radiation Sciences*. 54: 719–725.

RCR (Royal College of Radiologists). 2008. *On-target: Ensuring geometric accuracy in radiotherapy*. London: The Royal College of Radiologists.

RCR (Royal College of Radiologists). 2021. *On-target 2: updated guidance for image-guided radiotherapy*. London: The Royal College of Radiologists.

RFH (Sir H. N. Reliance Foundation Hospital). 2023. *India's first ETHOS adaptive radiotherapy with surface guided radiotherapy: Revolutionizing radiotherapy for personalized cancer treatment*. Available at: https://www.rfhospital.org/care-centres/our-technology/ethos#:~:text=We%20at%20Sir%20H%20N%20Reliance,of%20the%20ETHOS%20Adaptive%20Radiotherapy. (accessed on 28 November 2023).

Smith, G., Dunlop, A., Alexander, S., et al. 2023. Evaluation of therapeutic radiographer contouring for magnetic resonance image guided online adaptive prostate radiotherapy. *Radiotherapy and Oncology*. 180: 109457.

SOR (The Society of Radiographers). 2021. *The road to adaptive radiotherapy*. St Luke's Cancer Centre. Available at: https://www.sor.org/news/radiotherapy/the-road-to-adaptive-radiotherapy (accessed on 28 November 2023).

SOR (The Society of Radiographers). 2022. *The use of simulation in enhancing pre-registration education and training of therapeutic radiographers: Guidance document*. London: The Society of Radiographers.

Stewart-Lord, A., Swayne, T., Johnson, R., et al. 2022. The utilisation of VERT™ in the training of image-guided radiotherapy for therapeutic radiographers. *Journal of Radiotherapy in Practice*. 21: 88–91.

Teunissen, F., Willigenburg, T., Tree, A., et al. 2023. Magnetic resonance-guided adaptive radiation therapy for prostate cancer: the first results from the MOMENTUM study—An international registry for the evidence-based introduction of magnetic resonance-guided adaptive radiation therapy. *Practical Radiation Oncology*. 13: e261–269.

UoL (University of Liverpool). 2023. *MSC cancer care*. Available at https://www.liverpool.ac.uk/courses/2024/cancer-care-msc (accessed on 28 November 2023).

Varian (Varian – a Siemens Healthineers Company). 2023a. *CS502EU-Varian advanced imaging clinical school IGRT & motion management*. Available at: https://www.varian.com/en-gb/cs502eu-varian-advanced-imaging-clinical-school-igrt-motion-0 (accessed on 28 November 2023).

Varian (Varian – a Siemens Healthineers Company). 2023b. *CS508EU-Ethos adaptive clinical school*. Available at: https://www.varian.com/en-gb/course/cs508eu-ethos-adaptive-clinical-school (accessed on 28 November 2023).

Wallner, P., Jimenez, R. Davis, B., et al. 2023. Proton beam therapy training, experience, and assessment: Ready for prime time. *Practical Radiation Oncology*. 13: 286–288.

Wijeysingha, E., Chin, V., Lian, C. 2021. Utilising virtual environments for radiation therapy teaching and learning. *Journal of Medical Imaging and Radiation Sciences*. 52: S83–S95.

Zwart, L., Ong, F., ten Asbroek, L., et al. 2022. Cone-beam computed tomography-guided online adaptive radiotherapy is feasible for prostate cancer patients. *Physics and Imaging in Radiation Oncology*. 22: 98–103.

Index

Printed in the United States
by Baker & Taylor Publisher Services